Hamlyn all-colour paperbacks

95p

7

Ecology

Michael Allaby

illustra

The Tu

D0531082

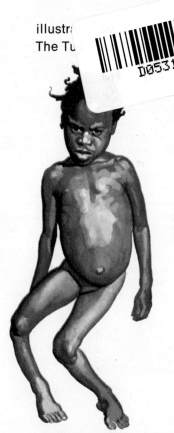

Hamlyn
London · New York · Sydney · Toronto

FOREWORD

Since the mid-1960s, the word 'ecology' has become fashionable. We read about the 'ecological impact' of proposed industrial, urban, or agricultural developments; of the 'ecological crisis' which may threaten our survival; we hear of people attempting to live 'ecologically'.

The word has acquired at least two meanings, one scientific, the other philosophical. In this book I have tried to explain what the science of ecology is and how the more popular interpretation of its name is derived from the stricter scientific definition.

Essentially, ecology is the study of communities and the relationships within and between them. Its principles are simple but its subject matter too complex to permit accurate prediction. The ecologist, concerned with observing adaptation and change, is studying evolution in action. From this rather obscure branch of the life sciences, it is but a small step logically, although a very large one scientifically, to expand ecology to the study of human communities. As a subject, 'human ecology' is not new, but growing awareness among scientists of disturbances to natural systems caused by human activity, led them to speculate about secondary effects such changes might have on man himself. From this point ecology became at once a science and a popular mass movement.

All of us can study ecology and most of us do, whenever we tend a garden, try to identify wild plants in the countryside, or watch the behaviour of wild animals. Through understanding ecological principles we may appreciate better the world and our place in it.   MA

The author would like to express his gratitude to John Barber, Sue Shires, and their colleagues for the very high quality of their illustrations, and to Neil Curtis for the quiet efficiency with which he organized the creation of this book – and for suggesting it in the first place.

Published by The Hamlyn Publishing Group Limited
London · New York · Sydney · Toronto
Astronaut House, Feltham, Middlesex, England

Phototypeset by Filmtype Services Limited, Scarborough, England
Colour separations by Metric Reproductions Limited,
Chelmsford, England
Printed in Spain by Mateu Cromo, Madrid

CONTENTS

INTRODUCTION

The all-togetherness of everything

Ecology has been called the study of the 'all-togetherness of everything'. It is the study of communities and the relationships that exist within them. As a science, it crosses the boundaries separating several disciplines. It is a branch of the biological, or life sciences, but the ecologist may find himself studying the earth sciences, such as geology or pedology, or the natural sciences, such as physics or chemistry. He will also need a good working knowledge of mathematics. Ecology is a general study of living organisms, their relationships to one another and to the planet on which they live, and of whole communities, from the smallest bacterial culture to the planet itself.

Not only will the ecologist study the communities as he finds them, but he will also watch communities, and the species within them, as the relationships among them develop. Ecology is very much concerned with evolution, therefore. Indeed, it is based very largely on the concept that species survive by adapting to environmental circumstances.

For this reason, ecology has acquired strong philosophical overtones. Man, after all, is a living organism and as such he is a proper subject for ecological study. He is a member of a community that includes species other than his own, and the ecologist is entitled to study the community as a whole and man's place within it. Like any other organism, man is affected by the micro-organisms, plants, and animals with which he must live, and he affects them. Yet here simple comparisons must end, because man's capacity to alter his environment far exceeds that of any other species. He intends to alter it to his own advantage, but whether this is to the advantage of other species, or of the community as a whole, raises questions of ethics and morality.

Our Earth is an oasis. In the infinitely large desert of space, and in a narrow band stretching from a few metres below its surface to a few thousand metres above it, it supports life in a multitude of forms, organized into communities. Ecology is the study of those communities.

Charles Darwin
(1809-82)

giant tortoise
(Testudo gigantea)

Malthus, Darwin, and Haeckel

In 1789, the English political economist, the Reverend Thomas Robert Malthus published his *Essay on the Principle of Population*, producing mathematical arguments to show that while populations increase geometrically, that is, by compound interest, the availability of the resources on which they depend increases only arithmetically, or by simple interest. The growth of populations must be checked by starvation, disease, and in the case of human populations, with which his *Essay* was concerned, by war and 'vice'.

This theory, like Darwin's, was used by opponents of social reform, who argued that to improve working and living conditions would be counter-productive.

Malthus' essay helped the English naturalist, Charles Darwin to clarify the ideas he expounded in *The Origin of Species*, first published in 1859. Darwin had sailed to South America on HMS *Beagle* and had visited the Galápagos, a group of islands on the Equator, 960 kilometres (km) from the nearest mainland. There he found unique species of animals, and in

Galápagos cactus finch
(Cactospiza heliobates)

flightless cormorant
(Nannopterum harrisi)

marine iguana
(Amblyrhynchus cristatus)

The name 'Galápagos' means 'tortoises'. and this giant tortoise may weigh 225 kilograms or more. Other unique Galápagos species include a four-eyed fish and a distinct penguin. Many tortoises have been killed for food, and feral domestic animals have damaged the habitat, reducing numbers of indigenous species.

particular a collection of finches that had adapted to different diets by becoming thirteen distinct species, not one of which existed elsewhere. He concluded that evolution proceeds by natural selection, the most adaptive members of a population being best able to survive pressures imposed by enemies and competitors and by the proximity of 'Malthusian limits'.

The word 'ecology' was coined in 1871 by the German biologist, Ernst Haeckel, who defined it as 'the study of the economy, of the household, of animal organisms'. The ideas of all three men were related to human society and Karl Marx and Friedrich Engels became embroiled in the ensuing political controversy.

During the early years of this century, however, the emphasis changed and ecology became one of the biological

7

sciences. *The Penguin Dictionary of Science* defines it as 'the study of the relation of plants and animals to their environment'. Man disappeared from the picture and with him went the social and political arguments of former times.

Ecology became a highly specialized branch of botany, and ecologists began to study in minute detail the composition of natural communities. Then entomologists and zoologists became involved as the study extended to animal species. Species were listed and the members of each one counted. Relationships between species were measured in terms of food supply and there developed the concept of a hierarchy of food producers and consumers. This, in turn, came to be seen as a process of energy and nutrient flow, whereby nutrients move through cycles from the soil through plants and animals and back to the soil. Energy is derived from the sun and is utilized by green plants and the animals that consume them. Ecology was moving close to thermodynamics and the mainstream of physics as attempts were made to quantify these cycles.

Ecology also moved closer to the earth sciences. The pattern of living species in a community is determined largely by climate and by the physical and chemical composition of the soil, and the modern ecologist should be trained in pedology, the science of soils, and in geology and geochemistry which study the origin of the rocks from which soil is derived.

New concepts began to appear, with new words coined to describe them: 'biosphere' for the envelope extending from the soil to a height of a few kilometres into the atmosphere within which life can occur; 'biomass' for the total mass of living organisms in any area; and 'ecological energetics' for the measurement of energy flows.

The work is exceedingly complex, because no two areas of land or water support exactly the same community. While much has been learned, some areas remain at the frontiers of knowledge. The ecology of soil populations, for example, is still fairly rudimentary. Techniques are still being developed and tested that will make possible the capture and accurate counting of soil organisms. The function of many of them remains obscure.

Ecology is a young science and one that is accumulating very rapidly the central core of information essential to any discipline. It is still concerned mainly with surveying and measuring, however. All science must proceed through an observational stage during which phenomena are recorded. From these collected data general concepts are formulated. Ecology is formulating concepts, but many of them are still tentative and exceptions can be found to most of its generalizations. Not until this theoretical basis has been established much more firmly will it be possible to move on to experiments designed to validate hypotheses. Only then will the general principles become verifiable. When that happens, ecology will become a predictive science and the ecologist will be able to forecast the result of events with a fair degree of accuracy.

A typical community of freshwater plants and animals that might be found in any village pond in Britain. In addition to the species that you can see, there are many more that are invisible to the naked eye, but which play a vital rôle in the welfare and stability of the whole.

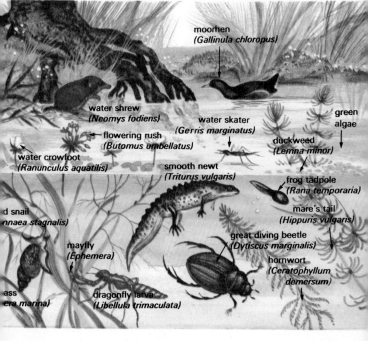

moorhen
(Gallinula chloropus)

water shrew
(Neomys fodiens)

water skater
(Gerris marginatus)

green algae

flowering rush
(Butomus umbellatus)

duckweed
(Lemna minor)

water crowfoot
(Ranunculus aquatilis)

smooth newt
(Triturus vulgaris)

frog tadpole
(Rana temporaria)

d snail
nnaea stagnalis)

mare's tail
(Hippuris vulgaris)

great diving beetle
(Dytiscus marginalis)

mayfly
(Ephemera)

hornwort
(Ceratophyllum demersum)

ass
era marina)

dragonfly larva
(Libellula trimaculata)

Human ecology

While ecology confined itself to the scientific study of environments, amateur naturalists became increasingly concerned at the loss of, and damage to, habitat caused by industry and the pressures of a larger and more mobile population. They became *conservationists*, seeking to protect areas of natural habitat that were especially vulnerable.

Some of them realized that damage to natural habitat might affect man. In 1962, Rachel Carson's *Silent Spring* was published in the United States of America, and in 1964 in Britain. Miss Carson was a professional biologist and her warning of the possible long-term effects of pesticide use encouraged other scientists to join what was by now a public debate. Biologists and ecologists began to examine in detail the impact of modern, industrial man on his environment.

In 1968, Paul R Ehrlich, a professor of biology, wrote *The Population Bomb*, which spelled out the biological implications of the current rate of growth of the human population. The Club of Rome, a private group of leading scientists, industrialists, and administrators, sponsored a study at the Massachusetts Institute of Technology, which was published in 1972 as *The Limits to Growth*. Earlier in the same year, *The Ecologist* magazine published 'A Blueprint for Survival'. They all concluded that by the depletion of the resources on which he depends, by pollution, and by the rate of growth of his own numbers, man faces a series of catastrophes early in the 2000s. Thus was born 'human ecology', although the term was not new, and a mass popular 'ecology movement'.

This has led to some confusion over the use of the term 'ecology'. Perhaps the most accurate description of those who aim to reform modern society so as to bring human behaviour into harmony with ecological principles, is conveyed in a phrase coined by Anne Chisholm, 'philosophers of the Earth' which she used as the title of her book describing the work and views of a number of such 'philosophers'.

Friends of the Earth, a group started in America and with active branches in several European countries, has campaigned for immediate reforms, using publicity to awaken public interest. It has also produced academic studies of particular problems.
Some campaign posters reproduced by permission of Friends of the Earth.

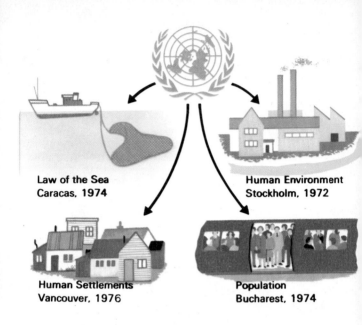

Law of the Sea
Caracas, 1974

Human Environment
Stockholm, 1972

Human Settlements
Vancouver, 1976

Population
Bucharest, 1974

Environmental topics now enter into any international conference concerned with human welfare or natural resources. In June, 1974, the United States and Soviet governments signed an agreement for a co-operative venture in saving wild lands in their two countries, and today no government can afford to ignore the effect on the environment of its activities.

Spaceship Earth

Adlai Stevenson first used the term 'spaceship Earth' and suggested that all living things on the Earth are its passengers. The image became popular as people came to see their planet as a whole, and life as dependent on biological and physical systems that are not immune to disruption. The pressure from some of the world's leading scientists, combined with the tangible evidence of gross environmental deterioration in their own countries, led to government involvement. The 1964 British Labour Government appointed a Minister with special responsibility for the environment, and the 1970 Conservative Government formed the Department of the Environment as the largest of its 'mega-ministries'. The government of the

United States and most European governments created a ministry of the environment, or its equivalent.

U Thant, then Secretary-General of the United Nations, made several speeches calling for international co-operation to arrest environmental deterioration, and agreement was reached to hold a conference on the human environment under UN auspices. Sweden was the host country and the Conference took place in Stockholm in 1972, attended by delegates from 113 countries including the People's Republic of China, but excluding the USSR and the east European countries. The Conference issued a Declaration on the Human Environment, established a global 'Earthwatch' monitoring system and created the UN Environmental Programme as a new agency led by Maurice Strong, organizer of the Conference and based, on the direction of the General Assembly, in Nairobi. Other inter-governmental agencies created their own environmental departments, and the Council of Europe, the Organization for Economic Co-operation and Development, and the European Economic Community all maintain a close interest in environmental matters.

Further UN conferences were planned to deal with particular aspects of the global environment. In 1974 the Law of the Sea Conference in Caracas discussed the exploitation of the seas and sea bed. Population growth was discussed in Bucharest in 1974, and in 1976 urban problems will be discussed in Vancouver at the Human Settlements Conference.

The speed with which national and international action has followed on scientific concern is an indication of the seriousness with which the problems are regarded.

In most industrial countries legislation has been passed to protect the environment, and although environmentalists, campaigning for improvement, may regard the actions taken as inadequate, nevertheless they indicate a heightened awareness. The principle according to which industries must pay to cure damage they cause is coming to be accepted and mining companies are often required to deposit money for the restoration of the landscape when their operations are completed. Efforts by conservationists, however, to secure a moratorium on the hunting of whales, some species of which are threatened with extinction, have failed.

THE BRICKS OF LIFE

The Earth was formed about 4500 million years ago. At first it was too hot to retain an atmosphere, but as the surface cooled, gases were discharged from volcanoes and held, to form an atmosphere of ammonia, hydrogen, methane, and water. With further cooling, the water condensed to form the oceans. Cosmic ultraviolet radiation and electrical discharges in the air provided the energy for chemical reactions in the seas that formed a 1 per cent solution of organic compounds. From these the first life was formed in the sea. The earliest living organisms probably appeared about 2700 million years ago.

This process has been simulated in the laboratory, with high electrical charges being passed through a 'primordial soup' composed of those simple compounds, dissolved in water, that are believed to have been present in the primitive oceans. Organic (carbon based) compounds can be made to combine and

The surface of the Earth as it may have appeared about 450 million years ago. Man is a newcomer. The mammals did not appear until about seventy million years ago, man's immediate forebears lived about twenty-five million years ago, and *Homo sapiens* probably appeared first about four million years ago.

reform into amino acids, the constituents of protein, which in turn is the basis of living tissue.

No life was possible on land until photosynthesizing marine plants had released sufficient oxygen into the atmosphere, where it was ionized by sunlight into ozone, which blanketed out the lethal ultraviolet radiation. Then, some 450 million years ago, the first life appeared on land.

The oxygen released photosynthetically by green plants is consumed again by bacteria when dead plant tissue decomposes. It is possible that the arrested decomposition of the great forests of the Carboniferous period, 345 to 280 million years ago, left our atmosphere with its present 23·2 per cent concentration of oxygen.

It is the same arrested process that has given us our deposits of coal, and a similar series of geological and climatic coincidences has left us with deposits of oil and natural gas, which are also derived from once-living organisms. For this reason, these are known colloquially as the 'fossil fuels'.

Violent changes in temperature, rains, winds, and volcanic activity broke and crushed surface rocks into fragments. It was in sheltered cracks that land life began. Soil formation began as soon as the first organisms were established on land. In

crevices by the sea shore, sheltered from rain and wind, single-celled plants and animals would be left by unusually high tides, blown from the caps of waves, or dried out and tossed by the wind from an exposed surface. Among them would have been some bacteria and microfungi able to extract the nutrients they needed from the rock itself. Our modern microfungus *Aspergillus niger* extracts potassium from stone, and several members of another group, the actinomycetes, take other elements. There are bacteria that will dissolve silicates and phosphorus. Organisms that can feed on purely mineral substances are called **autotrophs** and they are all single-celled.

Colonies of micro-organisms would form. As the first died

Eventually, these rocks will be covered with a layer of soil supporting vegetation, and it is the natural fate of all exposed stone surfaces to undergo this process. It has been observed on the walls of ancient buildings, even in the most arid parts of the world.

their remains would be consumed by others, and the colony would be protected from extremes of temperature, rain, and wind, by a jelly-like colloidal 'blanket'. To the naked eye the surface of the stone would appear rough or spongy as the organisms dissolved tiny holes into it. A thin layer of organic matter would accumulate, to be used by algae. These are visible blue-green or green streaks or patches. They are communities of single-celled plants, some of which contain chlorophyll and can form sugars by photosynthesis. At the edge of the sea, they may have included the *Laminaria,* brown seaweeds that can survive between the high and low tide lines.

Next would come lichens. These are algae that live in association with autotrophic microfungi, the fungi supplying minerals and the algae sugars. This kind of collaboration between different species is called **symbiosis** and the partners are **symbionts**.

We can guess that soil formation proceeded in this way although no-one observed it, because the process continues today, wherever a stone surface is exposed. Life builds up in layers, each providing food and shelter for the next. After the the lichens come mosses, simple plants with rudimentary roots. If you tear a piece of moss from an apparently bare wall you will find beneath it a very small amount of primitive soil. When enough soil has accumulated, larger plants can grow.

Even today, an undisturbed soil consists of layers, called 'horizons', each of which is clearly defined. The topsoil, or A horizon, contains most of the organic matter, partly decayed into humus and humic compounds, the subsoil contains much less organic matter, and at lower layers the soil becomes progressively more mineral. In most soils, but not all, the amount of organic matter is related directly to the soil's fertility, or its ability to sustain plants. Beyond this, soils vary very widely, according to their mineral composition, which is determined by the rock over which they have formed and climate and topography. Soil classification is extremely complex.

If you find an old wall, and part of its surface looks spongy, and it has blue or green stains and patches of a green crustation, with a little moss growing on it and perhaps a blade of grass, you will have seen all the stages in the formation of soil.

The chemistry of life

All life is related, and its chemistry is concerned with the arrangement of a limited number of elements into a vast number of compounds. These compounds, in turn, are used to construct sugars and starches (carbohydrates and celluloses), proteins, fats, and nucleic acids.

Altogether, sixteen elements are essential to all species. Carbon (C), hydrogen (H), and oxygen (O) are derived by life-forms from the air and water. Nitrogen (N), potassium (K), calcium (Ca), phosphorus (P), magnesium (Mg), and sulphur (S) are derived by plants from the soil and its micro-organisms, and are called the major plant nutrients. The minor plant nutrients, taken up in much smaller amounts, are iron (Fe), manganese (Mn), zinc (Zn), boron (B), copper (Cu), molybdenum (Mo), and cobalt (Co). Sodium (Na) is essential to some plants and animals, and chlorine (Cl) is necessary in very small amounts to all. Animals also need iodine (I) and selenium (Se).

Nutrient deficiencies cause disease and sometimes death. In human children, protein shortage retards development of nervous tissue and eventually of the whole body, when it may be diagnosed as marasmus or kwashiorkor. Carbohydrates provide energy and shortages may cause the body to 'burn' protein, producing clinical protein deficiencies.

Deficiencies of nitrogen, phosphorus, or potassium will cause discoloration and poor growth in plants. Deficiencies of calcium will inhibit growth in many species, and lack of boron will affect root development. In animals, calcium and boron are among the mineral nutrients. Calcium is required for development of teeth and bones. We derive most of our calcium from dairy produce, but it may also be obtained from fish bones and some cereals. We derive our chlorine and sodium from salt (NaCl). Vitamin D is produced by the action of sunlight on ergosterol in the body.

The pig suffers from lysine deficiency. Lysine is one of the amino acids from which proteins are made. The child on the left suffers from kwashiorkor, a protein deficiency disease, and the child on the right from rickets, caused by lack of vitamin D and sunlight. The corn is deficient in potassium.

The illustration of potassium starvation in corn is reproduced by permission of the Fertilizer Division of Fisons Limited.

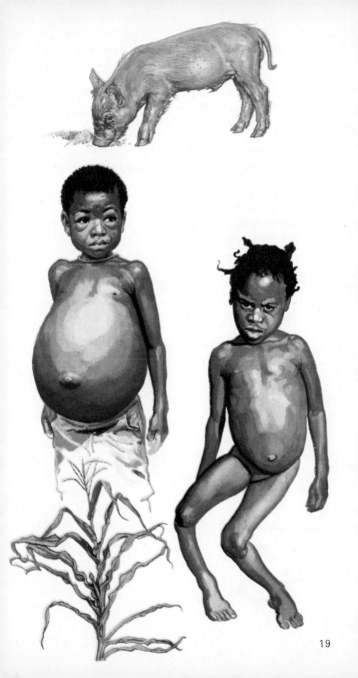

19

Carbon is one of the most common elements on Earth, and carbon atoms are able to form links with four other atoms. Where it is linked to less than four atoms, some of the bonds are double to bring the total to four. The human body consists of 65 per cent oxygen, 18 per cent carbon, and 10 per cent hydrogen by weight. All living tissue is composed mainly of these three elements. In man, the remaining 7 per cent is made up of nitrogen, calcium, phosphorus, potassium, sulphur, sodium, chlorine, magnesium, iron, manganese, copper, iodine, cobalt, and zinc.

The simplest carbon-hydrogen compound is methane, chemical formula CH_4 (one carbon atom linked by single bonds to four hydrogen atoms). Other simple compounds include such common substances as acetic acid (CH_3COOH),

The double and single lines indicate energy bonds that link atoms. If life occurs elsewhere in the universe, it is probable that its chemistry will be very similar to that on Earth, based on carbon, hydrogen, oxygen, and nitrogen.

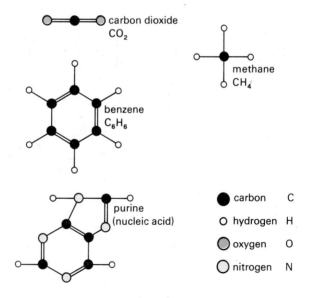

carbon dioxide
CO_2

methane
CH_4

benzene
C_6H_6

purine
(nucleic acid)

● carbon   C

○ hydrogen H

◔ oxygen   O

○ nitrogen   N

lactic acid ($CH_3CHOHCOOH$), and sugars, which vary in composition, but which can all be described by the formula $C_nH_{2n}O_n$.

More complex groupings are made possible when small molecules are linked. Palmitic acid, a constituent of fat, is $CH_3(CH_2)_{14}COOH$ and many molecules are much larger.

Life is possible because of this property of carbon to form the basis of giant molecules, and it is probable that if life exists elsewhere in the universe it will be dependent on similar chemistry.

The construction of protein and hence the reproduction of cells is possible because of the existence of the nucleic acids ribonucleic acid (RNA) and deoxyribonucleic acid (DNA). DNA acts as a template for the formation of RNA, and RNA for protein molecules. Therefore, it is through DNA and RNA that information is conveyed from one generation of cells to the next. In 1953, J Watson, F Crick, and M H F Wilkins received the Nobel Prize for physiology and medicine for their un-ravelling of the double-helix structure of the DNA molecule.

When light falls on the substance chlorophyll, the chloro-phyll molecule becomes 'excited'. The electrons of its con-stituent atoms vibrate so violently that the molecule can remain in existence only by releasing some of its surplus energy. This energy separates hydrogen from water. This is the first step in photosynthesis, by which green plants build sugars and starches from carbon dioxide and water, expressed as $6CO_2 + 6H_2O + \text{light energy} \rightarrow C_6H_{12}O_6 + 6O_2$.

Sugars then oxidize to carbon dioxide and water, giving up the energy derived originally from the sun. Energy is stored temporarily, however, by the synthesis of adenosine tri-phosphate (ATP) from adenosine diphosphate (ADP) and phosphate. The energy required for the reaction is released in the body when ATP reverts to ADP and phosphate.

Adenosine is formed from adenine and a sugar, ribose. It can then carry as many as three phosphates, each phosphate bond representing 8000 calories of energy. Each time a reaction occurs which liberates 8000 calories, it is possible for adenosine to form a phosphate bond. Whenever a cell requires 8000 calories, the addition of water will release one bond by hydrolysis.

The water cycle

Life began in the sea, and without water no life on land is possible. While living organisms can withstand extremes of temperature, they cannot survive in an arid climate. There is life in the Arctic and at the equator, but less in the hot deserts or close to the North and South Poles, where annual precipitation is very low.

Water is needed to convey nutrients to the roots of plants and through the bodies of animals, and the single cell is a semi-permeable envelope containing complex molecules in water. The mass of a cell is only slightly higher than that of an equivalent volume of water. Many chemical reactions occur in solution and while photosynthesis can take place in darkness, using stored energy, water must be present to give hydrogen.

If two solutions of mineral salts are separated by a cell wall, water will pass through the wall from the weaker to the stronger solution. The process is called osmosis and the difference in strength between the two solutions exerts an osmotic pressure. When the cells of land-based plants or animals encounter fresh water containing dissolved salts, water may flow in either direction according to the concentration of salts in the solutions. When they encounter salt water, however, the saline solution is always stronger than the solution within the cell and water flows out of the cell. This causes dehydration which can lead to death. This is why we die of thirst if we drink too much sea water, and why salted peanuts make us thirsty.

Land organisms have adapted to survive in a fresh-water environment, because evaporation abstracts pure water and leaves the salts behind.

Almost all the water on the planet is engaged in the hydrological cycle. Ninety-seven per cent of the world's water is in the oceans. Of the remaining 3 per cent, 98 per cent is in the icecaps. Were these to melt, the sea-level would rise by about 51 metres (m) and many lowland areas would be inundated. Each day about 875 cubic kilometres (km^3) of water evaporates from the seas and 775 km^3 falls on to the seas as precipitation. One hundred cubic kilometres is carried over the land, 160 km^3 evaporates from the land surface, and 260 km^3 falls as precipitation. The cycle is completed by the return of 100 km^3 to the seas through rivers.

The salinity of the seas has varied through the Earth's history. When the icecaps grow, for example, they do so mainly as a result of sea water freezing, but the precipitation they receive is held, frozen. In time they grow thicker as well as wider and the volume of sea water decreases. The salts are left behind when water is lost by evaporation, or distillation, so that the salinity of the sea will increase. Conversely, when the icecaps melt, the seas rise and their salinity decreases. Since the last Ice Age, the extent of the ice caps, and glaciers, has varied with climatic changes, and these have been traced for the last 1000 years by checking historical and harvest dates and records.

We use only a very small part of the total water at one point in the cycle, but the quantities can seem large. Six hundred litres (l) of water are needed to produce 1 kilogram (kg) of wheat.

For part of the Earth's history there has been no ice on its surface. At present, however, the icecaps are growing. If this process continues, sea-levels will fall and sea water will become more saline. There will be little effect — from this cause — on total fresh water precipitation.

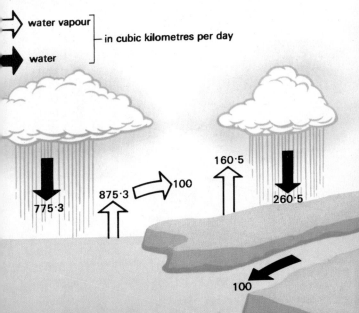

water vapour
water
— in cubic kilometres per day

775·3   875·3   100   160·5   260·5   100

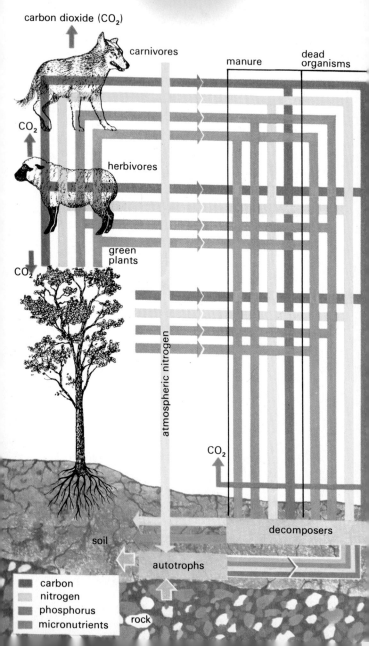

solar energy
for photosynthesis

carbon dioxide (CO_2)

carnivores

manure

dead organisms

CO_2

herbivores

green plants

atmospheric nitrogen

CO_2

decomposers

soil

autotrophs

carbon
nitrogen
phosphorus
micronutrients

rock

Nutrient cycles

In nature there is no waste. Nothing is created, nothing is lost. In the same way as water, nutrients are recycled constantly. Nutrients can and do move from one location to another, however, leaving local deficiencies.

The deficiencies are made good by autotrophs, which fix minerals from rocks and nitrogen from the air to form a nutrient reservoir in the soil. Part of the reserve of more soluble nutrients, such as nitrogen in the form of nitrates, is washed away by rainwater. The remainder is taken up by plants. Green plants take carbon from atmospheric carbon dioxide and hydrogen from water.

As plants complete their life cycles and die, small animals, insects and spiders, fungi and bacteria break them down into their constituent simple compounds, and they are added to the soil nutrient reservoir.

Plants are consumed by **herbivores** and herbivores are consumed by **carnivores**, which may also eat one another. All animals produce wastes, in the form of excrement and also their bodies when they die. All of this material returns to the soil reservoir, to be taken up again by plants.

Nutrients washed from the land find their way into rivers and lakes, where they join aquatic nutrient cycles and reservoirs to which nutrients are also brought as a result of the weathering of rocks. From rivers they may find their way to the sea. The seas are the ultimate 'sink' for most material removed from the land. It is not surprising that marine life is concentrated on continental shelves and along coastlines because the seas are so dependent on nutrients brought from the land. For much of their area, the deep oceans are rather sparsely populated, especially at their lower levels, few nutrients being found on the sea floor and such organisms as live there depending for their food on a nutrient reservoir above, rather than below them. But there is life at depths of almost 11 000 metres in the Marianas Trench of the western Pacific.

A cycling system like this might continue relatively unchanged for very long periods of time, because it is almost closed and a precise balance has been struck between gains and losses. It is highly stable.

Energy cycles

Many chemical reactions, all motion, and heat generation require a source of energy. Energy is required to power the great physical and biological cycles of the Earth.

Some of this energy is produced by chemical reactions and released by autotrophs, some is derived from the heat at the centre of the Earth, but most comes from the sun.

Each year the Earth receives about 15.3×10^8 ($10^8 = 100\ 000\ 000$) calories of solar energy at each square metre of surface ($cal/m^2/yr$). A calorie is *the energy required to raise the temperature of one gram (1 g) of water through one degree Celsius (1°C)*. Much of this energy is scattered by dust particles in the air or used to evaporate water. The amount reaching plants varies from place to place. In Britain it averages 2.5×10^8 $cal/m^2/yr$, in Michigan 4.7×10^8 and in Georgia (USA) 6.0×10^8. Ninety-five to 99 per cent of the energy received by plants is lost as reflected heat and heat of evaporation. The 1 to 5 per cent left is used in photosynthesis, which converts it to chemical energy to make it available to animals.

Agriculture replaces communities of wild plants with stands of crops that are useful to man. Beyond this, however, it aims to improve on original wild varieties so that domestic crop plants grow to a larger size than their ancestors. This is done by increasing the area of leaf so that more photosynthesis occurs.

The nutrient cycle may also be regarded as an energy cycle. There is a loss of energy from the system as it passes, in the form of food, from one range of organisms to the next. The diagram shows that of the total energy entering this system, plants make use of a net 2.15 per cent and pass on 0.82 per cent to the herbivores, which pass on 0.09 per cent to the carnivores and 0.005 per cent to the top carnivores. The energy available at each stage determines the theoretical limit to population size. There can be no more top carnivores than one twenty-thousandth of the energy entering the system will support.

The number of consumers may increase if the efficiency of photosynthesis increases. This may happen if among the plants there are some that capture significantly more than 2.15 per cent of incoming energy. In most temperate systems, however, plants capture less than 2 per cent of available energy.

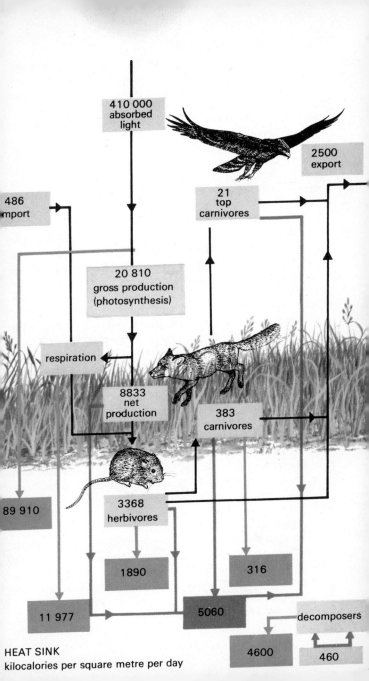

410 000
absorbed
light

2500
export

486
import

21
top
carnivores

20 810
gross production
(photosynthesis)

respiration

8833
net
production

383
carnivores

89 910

3368
herbivores

1890

316

11 977

5060

decomposers

4600

460

HEAT SINK
kilocalories per square metre per day

THE DISTRIBUTION OF FOOD

Plants are primary **producers** of food and all succeeding organisms are **consumers**. If a cow is grazing a field, the grass is the producer and the cow the consumer. If we follow the fate of a producer through the body of the herbivore that eats it and then through one or more carnivores, we see that the different species are related to one another rather like the links in a chain. In fact, they form a **food chain**.

Plants are also autotrophs, in that they are able to use nutrients in their simple mineral form. All consumers are **heterotrophs** and each stage in a food chain is called a **trophic level**.

From the energy flow diagram, we see that the producers utilize 1 to 5 per cent of the energy available (in this case 2·15 per cent) and at each subsequent level about 90 per cent is lost. The loss occurs because most of the food an animal eats is 'burned' to provide body energy. Very little is used to build new tissue, but it is on the accumulation of tissue at each trophic level that the next level depends.

We often forget just how important feeding is, not only for ourselves, but for all living things. It is important to remember, however, that finding the correct food in the correct amounts and proportions is one of the main stimuli behind the actions of all animals. Animals have very varied diets, whereas all plants depend upon the same basic nutrients.

No matter how efficient a biological system may be, it can never achieve 100 per cent efficiency. The First Law of Thermodynamics states that *energy can be changed from one form to another, but it can neither be created nor destroyed* (the Conservation of Energy); the Second Law states that *at every change a proportion will be lost as dispersed, unavailable heat*. An ecological 'rule of thumb' states that at each consumer level, about 10 per cent of the input energy is made available as consumable product.

In our own simple diagram, the plant passes on to the aphids about 2 per cent of the solar energy it receives. The aphids pass on to the ladybirds 10 per cent of that energy, and the ladybirds pass on 10 per cent of what they receive to the blackbird, which passes on 10 per cent of that to the cat.

We can add together the total masses of plants, ladybirds,

blackbirds, and cats and calculate what the total maximum mass of each may be. The total mass of living organisms considered in this way is called the **biomass**. If we give the plant biomass a value P, we would expect the mass of aphids not to exceed $\frac{P}{10}$, the ladybirds $\frac{P}{100}$, the blackbirds $\frac{P}{1000}$ and the cats $\frac{P}{10\ 000}$.

The concept of total biomass is often used to calculate the effect of environmental changes on a community. This can be misleading, however, because total biomass takes no account of the composition of the community and a pond dominated by one or two plant species and algae can seem equal to one with a complex, balanced community, for example.

We might expect from this that the biomass of the entire

Food passes from organism to organism along a food chain. At each stage, roughly 10 per cent of the energy received by the organism is passed to the organism that consumes it. Obviously, therefore, the ladybird must eat large numbers of aphids, the blackbird large numbers of insects, and the cat must eat many birds.

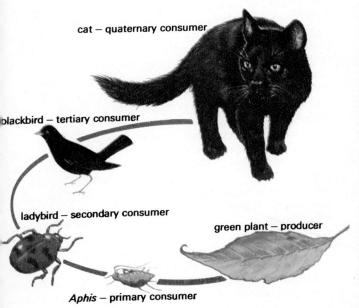

cat — quaternary consumer

blackbird — tertiary consumer

ladybird — secondary consumer

green plant — producer

Aphis — primary consumer

system would be $P + \frac{P}{10} + \frac{P}{100} + \frac{P}{1000} + \frac{P}{10\,000}$. This would account for only part of the system, however. We have omitted all the organisms that exist prior to the plant, and on which it is dependent. The plant is not independent of soil although it is autotrophic. It can obtain carbon from the air and hydrogen from water, and its roots can take up simple mineral salts from the aqueous solution in the soil, but the salts themselves must be provided by other trophic levels. The plant cannot fix atmospheric nitrogen unaided, nor many of the other nutrients it requires.

We might expect the biomass below the producer level to be the largest of all because the biomass at each trophic level decreases as we move further away from the primary producer. The roots of a plant may occupy more space than the part of the plant that is above ground, and because the region immediately surrounding the plant root system, the **rhizo-sphere**, supports a complex community of micro-organisms, our supposition is almost certainly correct. Roots grow and die back each year like the exposed parts of the plant. The total amount of new growth in a year is called the annual production, and measurements have shown that root production amounts to between 15 and 20 per cent of total production. Trees are located towards the lower end of the range and grasses towards the higher end. It is known that the total biomass of the soil generally exceeds that of the community above the soil and it is believed that micro-organisms may account for as much as 90 per cent of the energy flow through the whole system.

Beyond this, accurate measurement is notoriously difficult. Soil populations are easily disturbed. The larger animals, such as earthworms, will move away from a disturbance, so that counts of their numbers based on samplings may be too low, while microbial activity increases sharply if the soil is moved and additional air introduced. Very little is known about the rôle of the micro-organisms in general or of particular species. We do have some idea of the kind of numbers involved, however. Population counts made in a forest near Paris revealed in the top 100 mm of a 1 m^2 area more than thirty million nematodes, which are small worms, 45 000 mites of one species alone, and while they did not count the springtails (wingless insects with elastic tail-like appendages) they examined, British workers have done so and have found more than 2000 in a similar block of soil.

The study of energy and nutrient flows through soil populations is now the subject of intensive research under the International Biological Programme, which began in 1964 and ended in 1974. Of the seven sections of the IBP, one of the largest was that concerned with terrestrial productivity, which includes the study of soil organisms. One of its tasks was to standardize the methods used by biologists throughout the world so that their results conform to similar criteria. Without this it is difficult to compare findings and build an overall picture.

It is possible to bypass the soil systems such as the one illustrated here by supplying plants with a nutrient solution. This process is called **hydroponics**. The nutrients must be made and applied and the wastes removed by man, who in effect assumes part of the rôle of the soil organisms.

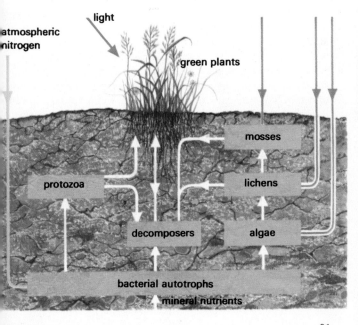

Decomposers

The vast and complex community of soil organisms that makes nutrients available to plants consists almost entirely of heterotrophic plants and animals, the decomposers. The autotrophic micro-organisms are probably much smaller in biomass and the most important autotrophs are green plants themselves.

Decomposition begins the moment organic litter is deposited on the floor. Experiments in which oak and beech leaves, sealed in nylon mesh bags, were buried 250 mm deep in freshly turned soil, showed that after nine months, more than 90 per cent of the leaf mass had disappeared in those bags with mesh which permitted earthworms to enter, but that mesh too small for earthworms slowed down decomposition by two-thirds. The worms reduce the size of the leaves, leaving tiny fragments as well as their own excrement, for other, smaller organisms. French workers have studied the fate of a single pine needle 60 mm $\times$ 1 mm $\times$ 0·5 mm in size. Oribatids, minute mites, broke the needle into thirty million cubes, then nematodes broke these into 3000 million cubes measuring $(0·0001 \text{ mm})^3$. The surface area of the original needle was 80 mm^2. The surface area of the cubes was 1 800 000 mm^2. It is the surface area that is attacked by bacteria and fungi.

The decomposition of dead animals may begin with the scavengers, including such birds and animals as cats, dogs, and crows which eat carrion when it is available but do not depend on it. Fly larvae, earthworms, woodlice, millipedes, and the larger fungi, such as toadstools, attack the smaller fragments of animal detritus and all plant wastes. Then the mites, springtails, and nematodes take over. Finally, it is the turn of the microfungi, protozoans, and bacteria. When they are done, the simple compounds which they add to the soil solution are taken up by plants and the cycle begins again.

It is not only organic materials that are subject to biological degradation. The autotrophic organisms will attack any suitable material. Experiments in Germany to discover the cause of the deterioration of optical lenses during storage and in long sea voyages led to the identification of microfungi that will break down glass. Subsequent investigations have shown that this does occur when fragments of glass reach the soil, usually in composted urban refuse from which glass can be separated only with great difficulty.

It is impossible, then to estimate the total biomass of our simple food chain without including organisms that are engaged in recycling materials from the top of the chain back to the bottom.

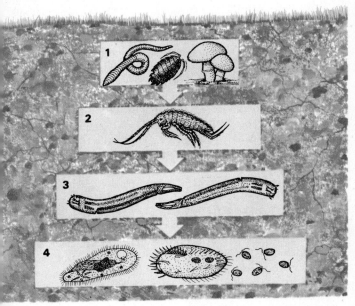

The living organisms in the top few centimetres of a hectare of fertile soil may weigh more than a cow grazing the pasture growing on that hectare. The organisms involved directly in the nutrient cycle support their own populations of predators, parasites, and symbionts. **Key:** 1 worms, woodlice, toadstools etc. 2 mites, springtails etc. 3 nematodes. 4 protozoa, microfungi, bacteria.

If the soil is sustaining the maximum producer-consumer biomass of which it is capable, the residence time of nutrients in the soil will be short, and there will be little or no storage of nutrient energy. The speed with which recycling takes place depends on the input of solar energy, and a tropical soil recycles faster than a cold one.

Chains or webs?

Our simple food chain is too simple to correspond to a situation in the real world. A blackbird will eat ladybirds if it comes across them, but normally it feeds on the ground and its diet includes earthworms, insects, grubs, seeds, and fruits. Ladybirds are of no great importance to it. If we take into account all the items in its diet, and trace the source of all these items, and fit the blackbird into a picture that includes the cat that hunts it, our simple food chain begins to look much more like a web. To complicate matters still further, many species change their eating habits from season to season.

Food webs can become extremely complex. That of the tawny owl is fairly simple. The tawny owl is the most common British owl. It lives in woodland, open land, and even in cities, wherever there is a large tree with a hole in which it can nest.

From December to February, owls may feed on wood mice and voles. From May to August they eat moles, young rabbits, earthworms, and beetles. Mice and voles form about half their total diet.

The wood mouse, which is nocturnal like the owl, eats mainly seeds but it will eat insects, especially in spring. The bank vole may be active by day or night. It feeds on dead or living vegetable matter. Moles eat insects and earthworms, and earthworms are decomposers. The beetles eaten by the owl include cockchafers which are herbivorous, and ground beetles which eat other insects. Both wood mice and voles breed from April to October, but voles may be more plentiful in winter.

Thus, the food web is subject to seasonal variations due to the hibernation of some species and the breeding habits of others. The web is much more dynamic than it may appear to be in the diagram. This dynamism might be shown if the size of connecting lines also indicated quantity of supply.

The entire web depends on the relationships between the

In real life, of course, many more organisms than are shown in this simple web may be involved. The tawny owl is not the only predator of small rodents, and insects and beetles are eaten by other birds. The food web for a complete area, rather than a single top predator, is much more complex.

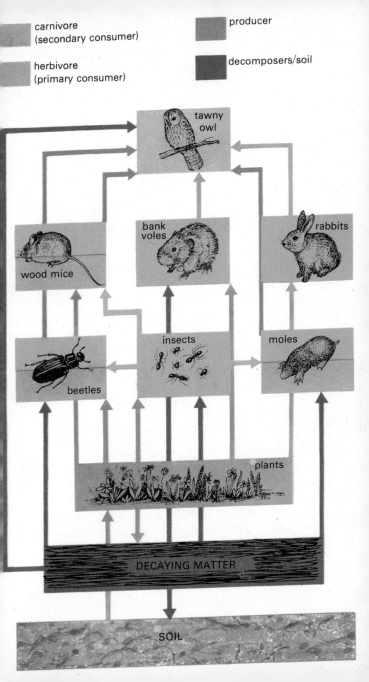

carnivore
(secondary consumer)

producer

herbivore
(primary consumer)

decomposers/soil

tawny owl

bank voles

rabbits

wood mice

beetles

insects

moles

plants

DECAYING MATTER

SOIL

organisms within it, so that it is clear that the removal of certain organisms will create vacancies which others may enter the web to fill, while the removal of others may disrupt the web totally.

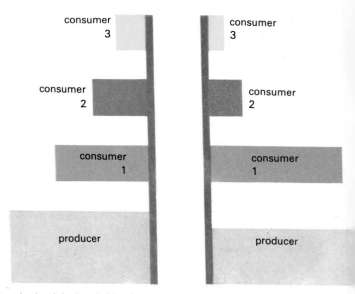

In the right-hand side of the diagram the population of secondary consumers has been reduced. The population of herbivores (consumer 1) increases, and the population of producers is reduced, as is the population of consumer 3. Overgrazing may reduce the herbivore population later. In other words, if the situation continues, the increased herbivores will reduce the producers until the herbivores decrease.

Our tawny owl web might be taken back further to include the soil organisms. If they were removed from the web, the plants could not survive. If there were no plant producers, there could be no primary consumers. Without herbivores there could be no carnivores. The entire web depends on the soil population.

Let us instead, remove the wood mice. They eat seeds and insects, so that another creature may enter to take advantage of this food supply, to fill the 'niche'. Provided nesting accommodation and materials were available, a bird might do so. The bird feeds by day so that the owl will not eat it, relying more on voles. This will cause a decrease in the vole population. This could be critical for the owl in winter. If it survived, it could do so only with a larger hunting territory. The absence of voles would lead to an accumulation of decaying vegetable matter. This would trigger an increase in the earthworm and beetle population, but the potential represented by the food supply would not be realized because the hungry owl, now assisted during daytime by the bird, would keep them in check. So, in the end, our web might contain fewer owls, a bird species in place of the mice, fewer voles, rather more worms and beetles and a larger accumulation of dead vegetation.

Over the last few years, the snowy owl (*Nyctea scandiaca*) has been nesting in Scotland. It is usually considered to be a more northerly species, but it is just one of a number of birds to move south as the climate has become cooler. One year, the male of a pair of snowy owls that had nested in the same place for several years appeared with two mates, which established two nests. One nest failed, however, because the male was unable to feed both families, and one clutch of eggs failed to hatch. In this way, too, the size of a population may be limited. (*Note:* the tawny owl is *Strix aluco*)

The removal from a web of a single organism may produce effects that seem disproportionately large, although, of course, it may be replaced by an organism that fits exactly the rôle vacated, not only eating the food supply, but feeding the same predators. In such a case no significant change may occur.

Obviously, the removal of an entire trophic level will have a disastrous effect. At producer level or below it will mean the collapse of the entire web. We have seen the effect of removing one primary consumer. If we were to remove all the primary consumers, nothing above that level could survive. Below that level, with no herbivores to graze them, the plants would increase in size and number, although their capacity to do so would be limited by the availability of nutrient in the soil.

The removal of organisms need not effect biomass. As the web adjusts, the final total biomass may remain static.

The top carnivores

What would happen, then, if we were to remove the owl from the top of the web? Obviously, the mice, vole, and mole populations would increase. This state of affairs might not endure long, however. The dead vegetation that forms part of the vole diet cannot exceed a fixed proportion of the annual plant production. More voles would deplete this store more quickly and voles would have to rely more on live plants. Presumably, these are being grazed by primary consumers, with which the voles would have to compete. If they were unsuccessful, the exhaustion of the dead vegetation could leave a larger number of animals with a diminished food source. This could bring about the collapse of the vole population. The vole might disappear altogether, or its numbers might re-stabilize at a lower level than when voles were being eaten by owls. A similar fate might await the mice and moles.

We are assuming that the community must feed in one small, fixed area. A population increase that exceeds the carrying capacity of the territory may equally lead to a mass migration. Therefore, the disruption of one web may lead to the disruption of several other webs nearby.

Top carnivores play a vital rôle in regulating the webs of which they are part. In Sierra Leone, farmers found that the small vervet monkeys were increasing in number and causing serious damage to crops. Investigation showed that it was the creation of new fields that caused the problem. The habitat of the leopards was being destroyed, and leopard numbers were declining. Leopards are at the top of the web that includes vervet monkeys.

Insect populations sometimes experience cycles of population growth and collapse. A decline in an herbivorous species causes a decline in its predators and the removal of predator pressure allows the prey to increase again in the time lag between prey and predator expansion.

Most of the top predators are carnivorous. Man is omnivorous and eats plants as well as animals. The increase in his own numbers has been made possible by agriculture, which increases the production of species he is able to eat, at the expense of his competitors.

39

The praying mantis, *Pseudocreobotra wahlbergi*, is a predatory insect not found in Britain, but common in North America where it is sold to farmers and horticulturists for pest control. The ladybird is sold in the same way, and in Britain growers can buy species of predatory mites.

Predators and their prey

Let us imagine now what would happen if a year of exceptionally favourable weather produced a substantial increase in production. The primary consumers would increase in numbers and the carnivores would also increase to eat them. A food source will always attract a consumer. Once the food source is exhausted, numbers return to their previous level. The balance of nature is maintained at all times, but the balance is dynamic, constantly changing. The population of primary consumers is regulated by the availability of food and by the secondary consumers. The population of secondary consumers is regulated by the number of primary consumers and the activity of tertiary consumers.

There are exceptions, however. The desert locust, *Schistocerca gregaria*, is two insects in one. Normally it lives like any other primary consumer in its rather barren web. Plant production is subject to wide seasonal variation, increasing after the rains and then dying back, so that the locusts are constantly expanding and then being crowded together. From time to time, when the locusts come together, they alter their behaviour. The solitary insect becomes gregarious. Its body colour and its entire manner change, so that for many years it was thought to be a different species. It breeds rapidly and then migrates, in immense swarms, in search of food. It has no predator to regulate its numbers, except perhaps the poor Arab farmers, who are reduced to eating the insects after the insects have eaten their crops. Ironically, it may be the farmers who are responsible for at least some of the depredations they suffer. Irrigation provides year-round food supplies for the solitary insects, while crop stands offer sustenance to migrating gregarious swarms.

Pests are not found in nature, of course, where checks and balances work to maintain stability. Farmers create pests by offering large sources of food to primary consumers many of which are so specialized that they will eat only one kind of plant. Farming can maintain its own kind of ecological stability, but when it does not the results may seem bizarre. Industrial development in the north of Italy attracted farm workers from the south. Many fruit farmers left so hurriedly that they did not wait to harvest their crop. The fruit fell from the trees and decayed. This attracted small rodents (primary consumers) and the rodents attracted secondary consumers, in this case snakes. The infestation by venomous snakes became so severe that official warnings were given to visitors. A web that is not disturbed from outside, however, will remain indefinitely in a state of dynamic equilibrium.

Attempts at outside control may prove counter-productive. Insecticide use tends to breed resistant pest populations more rapidly than resistant predator populations, because pest populations are larger. The removal of one insect species or weed may allow a rival species to flourish and the rival may be more difficult to control. If weed control is too efficient, herbivorous insects may be driven to the crop as the only available food supply, so exchanging a weed problem for a pest problem. Modern methods of biological control are usually more efficient than pesticides.

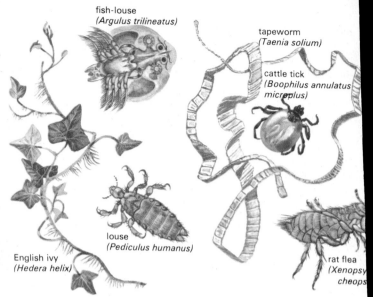

fish-louse
(Argulus trilineatus)

tapeworm
(Taenia solium)

cattle tick
*(Boophilus annulatus
microplus)*

louse
(Pediculus humanus)

English ivy
(Hedera helix)

rat flea
*(Xenopsy
cheops*

Species of common parasites. The parasite's host may not die, but
its growth and health is impaired. Blood-sucking parasites such as
fleas and ticks may carry secondary infections which enter through
the wound they make and which may be fatal.

PARASITES AND SYMBIONTS

Every species within a food web has a relationship with every
other species in the same web, regardless of whether it eats
or is eaten by that species. Yet so far we have considered only
the simplest of relationships, those between the eater and the
eaten.

There are two more groups of species that have a different
kind of relationship, the **parasites** and the **symbionts**. Both
are as dependent on their host or partner as the hunter is on
its prey, but very often the food source is not killed and it
may benefit. An organism that subsists by taking nutrients
from another organism, often to its detriment and never to its
advantage, is a parasite. An organism that cannot survive
independently of another organism, but that lives in partner-
ship with it to their mutual benefit, is a symbiont and the

relationship is called **symbiosis**. An organism that parasitizes a parasite is a hyperparasite and there are hyper-hyperparasites. The relationships are easy to define, but not always so simple to recognize in the field.

Consider, for example, the case of some central European oak and ash forests, in which oak and ash seedlings cannot survive beneath the shade of the mature trees for lack of light. Black beech seedlings can survive because they need less light. The trees compete with one another for soil and carbon dioxide, but between them they regulate the humidity to their mutual benefit, so that they may be symbionts. On the other hand, the beech seedlings depend on the shade provided by the larger trees, and they may be parasites.

Other examples are clearer. Hermit crabs, which live in deserted snail shells, carry a sea anemone (*Actinia*) on their backs. The anemone feeds on morsels from the crab's meals and is transported to feeding sites. At the same time it protects the crab with its poisonous tentacles. This is symbiosis.

Symbiosis may be subdivided further into **mutualism** and **commensalism**. The hermit crab and the sea anemone have a mutualistic relationship, in which both benefit. In other species, structure and behaviour may be modified greatly to gain the maximum advantage from the relationship.

Commensalism is a relationship between two organisms of benefit to one but irrelevant to the welfare of the other. Some shellfish live in holes made by other species, for example, and in the mammalian gut there are amoeba which subsist on the contents of the gut but have no discernible effect on the host. Pierre van Beneden (1876) said that a commensal 'requires from his neighbour a simple place on board his vessel, and does not partake of his provisions. . . . All he desires is a home and his friend's superfluities'.

Parasites are usually much smaller than their hosts, but insects which parasitize other insects of much the same size are called **parasitoids**. The small wasps that lay their eggs on or close to the host so that the larvae may feed on the host's body from the inside are parasitoids. In these cases the host invariably dies. The ivy that lacks a rigid stem of its own and so uses the trunk of a tree to climb toward the sunlight is a parasite.

Boophilus annulatus microplus has created severe economic problems for Australian farmers. It was introduced in the 1800s, probably from Asia, and is now resistant to most insecticides. The cattle egret and mynah, which might have controlled it, are absent.

Animal parasites may have complex life cycles, with more than one host. The cattle tick, *Boophilus annulatus microplus*, has only one. It begins life among grass where it hatches as a 'seed tick', attaches itself to a leaf, and then to any animal that brushes against it. On the animal's skin it moults into a nymph and feeds on its host's blood for about ten days, by which time it is adult, mating takes place and then the female engorges herself and falls from the host to lay her eggs in the grass.

The cattle egret and the Indian mynah, two birds about the size of blackbirds, live symbiotically with cattle, eating insect parasites and so freeing the animals from them.

The liver fluke is a two-host parasite. An internal worm, shaped like a leaf, it is common in wet lowlands which also favour its first host, a mud snail. Adult flukes lay their eggs in the body of their mammalian host. The eggs leave the body with the droppings and hatch into larvae which seek out the snails and develop inside their bodies. About ten weeks later they emerge as tiny tadpole-like creatures, attach themselves to damp grass and form a protective outer covering. In this, cyst, form, about the size of a pinhead, they are eaten by mammals. The covering dissolves in the stomach and the young flukes bore through the liver to the bile ducts, where they mature.

The common housefly, *Musca domestica*, is a free-living parasite. It lives on many liquids, including blood, but it possesses no mouth parts capable of piercing skin and so it is not a blood sucker, as are many fleas, lice, mites, and ticks.

Many free-living insects lay their eggs in decaying meat. This may have led to the laying of eggs in open wounds, and to true parasitism. Strike, a disease of sheep that is sometimes fatal, is caused by a carrion-feeding fly that has become

The female tick detaches itself from the host and falls to the grass to lay its eggs and die. The eggs hatch into 'seed ticks' that cling to plants, transferring themselves to the skins of animals that brush against them. There they mature as parasites.

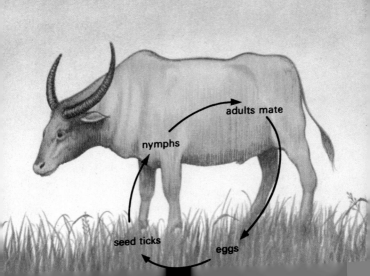

adults mate

nymphs

seed ticks

eggs

parasitic only recently. It lays its eggs on fleece contaminated by faeces or urine, and the maggots invade the flesh.

In purely numerical terms, both with regard to the range of species and to the numbers of individuals within each, the majority of parasites and symbionts are microbial.

Many diseases are associated with rapid increases in the populations of bacterial parasites. In man, food poisoning may be caused by *Salmonella* bacteria, one of the most common being *Salmonella typhimurium*. *Escherichia coli* is fairly harmless, but it is a close relative of the typhoid bacillus. Anthrax, brucella, staphylococcus, and streptococcus are all disease-causing bacteria that may parasitize several mammals, including man.

Parasites themselves may have commensal relationships. Skin parasites that subsist on the blood of their host usually cause little trouble in themselves, but the wound they make and the anticoagulant they inject, create a habitat in which micro-organisms may flourish. Thus, skin parasites, and especially blood-sucking species, are often disease carriers. Certain species of mosquito carry malaria, the tsetse fly carries sleeping sickness, and ticks carry a variety of diseases known collectively as tick fevers. The South American vampire bat is rightly feared: the bat population is largely rabid and transmits rabies to the host.

The mammalian gut supports complex communities of microflora and microfauna. Most are harmless parasites (commensals), but some are symbionts with activities that are essential to the well-being of their partner. A main function of the large intestine is to accommodate these micropopulations. The gut of an infant is invaded by bacteria within a few days of birth and the colonies remain there for the rest of the animal's life. They synthesize several vitamins from the digested food matter passing through the gut. Most mammals are unable to manufacture for themselves all the vitamins they need. In man, all of the B group and vitamin K are made available by the gut flora. When antibiotics are administered therapeutically, some of the gut flora may be killed, and alternative sources of B vitamins are usually given to the patient at the same time as the drug.

Nor can animals break down tough celluloses unaided.

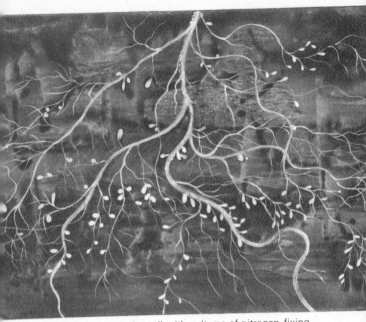

It is possible to inoculate the soil with cultures of nitrogen-fixing organisms, and in years to come this technique may provide an alternative to expensive nitrogen fertilizer. For centuries farmers have grown legumes in order to improve the fertility of their soils.

Ruminants are able to eat tough materials such as grass because of the micropopulation in their complex stomachs. Bacteria are able to use protein in forms that are not available to the mammal. The bacteria are eaten by protozoa, of the group *Entodiniomorpha*, and eventually the protein is released. The ciliate (having short hairs on the surface for locomotion and so on) protozoa are doubly symbiotic. Not only do they benefit the animal in the stomach of which they live, but they themselves live and work as groups, clustering together to form a 'boat' that is propelled by the cilia on the outside, some of which form a 'rudder'.

Unlike most plants, which remove available nitrogen from the soil, legumes add nitrogen. Attached to their roots are nodules, each of which contains a colony of nitrogen-fixing bacteria. When the plant dies, or is removed, the roots are

47

left behind, with their nodules, increasing the soil nitrogen content. Where legumes grow for the first time, the requisite bacteria may be absent, in which case the legume behaves like any other plant.

The nitrogen-fixing bacteria associated with legumes contribute to the welfare of the plant community as a whole. Were they absent, the nutrient status of the soil would be lower and fewer plants could be supported. If the gut of mammals lacked its flora and fauna, the animals could not survive. Without the parasite-eating symbionts, such as the cattle egret, the health of their partners would be poorer and the species could have fewer members.

The effect on a community of the removal of certain symbionts may be analogous to the removal of a trophic level. In the absence of the legume nodules, for example, there will be fewer legumes and a generally poorer plant cover. Consequently, there can be fewer herbivores. The legumes

All the larger, more complex organisms — those you can see — depend on symbionts in one way or another. All vertebrates have populations of bacteria contributing to their welfare. Similarly, they all have parasites. In birds and mammals, grooming helps remove skin parasites

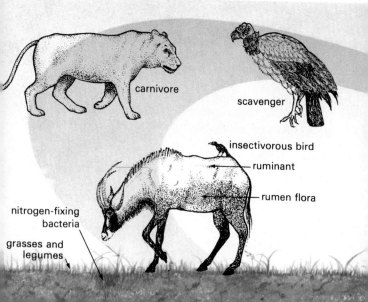

carnivore

scavenger

insectivorous bird
ruminant
rumen flora

nitrogen-fixing bacteria

grasses and legumes

are particularly nutritious plants, however, rich in protein. Therefore, the effect on the populations of consumers may be larger than a simple change in the structure of the plant community would suggest. In the community we have examined, there might be a reduction in the mouse population, which would be reflected in a reduced owl population.

Parasites, too, perform a useful rôle with regard to the community as a whole. If all the members of a species are equally available for parasite attack, it is the weaker among them that will be more susceptible and that, once attacked, will be most seriously affected. They will suffer a higher mortality rate from disease than their more robust rivals and their reproductive performance will be diminished. Thus, not only are the stronger better able to withstand attack, the attack will have the effect of removing rivals for mates and competitors for food. It is they who will breed, ensuring a healthy population.

If there are no nitrogen-fixing bacteria, the pasture will be poor and there will be fewer consumers; without rumen flora there can be no ruminants, and there will be fewer consumers; without insectivorous birds, there will be fewer ruminants and rather fewer consumers.

no nitrogen-fixing bacteria	no rumen flora	no insectivorous birds
fewer scavengers	fewer scavengers	fewer scavengers
fewer carnivores	fewer carnivores	fewer carnivores
fewer herbivores	no ruminants	diseased ruminants
poor pasture	overgrown pasture	overgrown pasture

The biotic pyramid

We saw earlier that when nutrient, or energy, passes from one trophic level to another there is a loss, and that the biomass at each level must be less than the biomass at the preceding level. We are in a position now to represent our community in a diagram that shows the biomass at each level. We must begin with the producers, including those soil micro-organisms that make nutrients available to plants, and the microscopic autotrophs. Above them come the plant producers, above them the herbivores, or primary consumers, then the carnivores or secondary consumers and, if there are any, the tertiary and quaternary consumers. We see that if we represent each level by a block with an area proportional to the biomass at that level, the figure represents a pyramid or, more correctly, a stepped pyramid, a ziggurat. It is called the **biotic pyramid** and it shows very dramatically the dependence of each level on those below. It shows, too, why it is unusual to find consumers beyond the fourth level. The reduction in biomass at each level is so great that by the time the fifth consumer level is reached the food supply is unlikely to support a biomass consisting of enough individuals to form a breeding population.

If, now, we add the decomposers to the pyramid, we find we have drawn an inverted pyramid below the first. To the decomposers, the organic detritus represents a food supply that is passed from trophic level to level as molecules are broken down. Again, at each level there is a loss and each level must have a smaller biomass than the level that preceded it. The same system operates, but in reverse.

The symbolism of the pyramid must not be taken too far, however. It is convenient to show trophic levels arranged in this form, but in real life they exist side by side and in no sense is any one organism, species or level any more or less important than any other. The hierarchical structure carries no political overtones! Nor does the fact that one organism is prey to another guarantee that it will be eaten should the two meet. Predators ignore food unless hungry, and it is in the interest of the predator to ensure the survival of its prey population.

The pyramid is useful as an illustration of the quantitative relationships between trophic levels, but it suggests a hierarchical structure far more rigid than is found in nature. At the

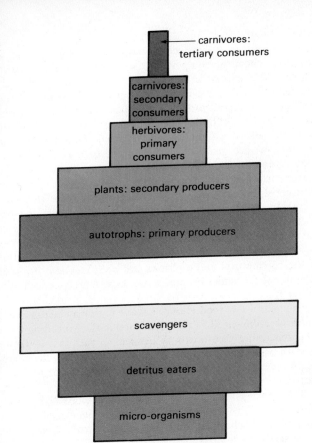

The biotic pyramid is a convenient way of illustrating the dependence of each trophic level on the level below it. Very roughly, each consumer block should be one-tenth the area of the one before, the first being one-tenth the area of the plant block.

secondary consumer level, for example, omnivores may subsist directly on two lower levels and, indeed, may be able to survive the removal of herbivorous species by assuming their rôle. Nor should we forget that nutrient flows cyclically through natural communities, not in a linear fashion.

51

Succession

Communities develop **successionally**. A parent rock of a particular chemical composition, weathered to a particular degree, under particular conditions of temperature, sunshine, and rainfall, will provide a food source and shelter for the first range of organisms. Their modification of their environment, together with the accumulation of their own waste products, afford a food supply for the next group. So the presence of each group of organisms prepares a niche for the next. As soon as plants appear there is food for herbivores and, if there is a surplus, for omnivores. These provide food for carnivores. At each stage new groups of parasites and symbionts appear to exploit a host or partner species, and opportunists arrive, adapted to survive in a wide range of environmental conditions and on a diverse diet. Eventually no more consumers can be supported by the total production of the community, and a **climax** has been reached. If a food supply exists, sooner or later a species will arrive or adapt to utilize it.

The quantity of food available at any level may be calculated as roughly one-third of the population at the preceding level. The remaining two-thirds is needed to sustain population growth and reproduction. What is eaten is surplus to the requirement of the prey population. Were it not so, over-grazing would be the rule.

This process can be observed where woodland adjoins untended land. Grasses and herbs are overtaken by low shrubs and bushes and these give way in turn to trees, grown from seeds carried from the wood.

A successional community is one in which food or shelter remains unutilized. A climax community has no food and no shelter to sustain new arrivals. Surplus food is stored in the soil and a successional community is likely to have a soil richer in nutrients than a climax community. If a climax community is simplified, it becomes successional, moving towards a new climax.

This picture is familiar wherever woodland is spreading to adjoining areas, but it also occurs far from the nearest woodland. Bare earth is colonized by grasses, herbs, and flowers, followed by shrubs, followed by trees, seeds being carried by the wind and by birds.

Where soils retain moisture better, a similar climate will support a far wider range of plant species. You might see a community such as this, of scrub and a few trees with a good cover of ground vegetation, within a few metres of an area covered with blown sand.

The kind of community that exists in any area will depend in the first instance on the chemical composition of the parent rock and then on the climate. Obviously, a granite mountainside, from which all soil and plants are swept by rain and wind before they can establish themselves, will support only simple communities. There will be no trees or large plants and because productivity is low there will be little to attract large grazing animals. Such animal species as do survive must feed from a very wide area. In Britain, for example, farmers can feed a cow on 0·4 hectares (ha) of land. In most of the world a cow needs 1600 ha or more. Little plant growth can take place below about 5 °C, and communities close to the snow-line will be relatively simple, as will desert communities that must survive with very little water. During the arctic summer, and when it rains in the desert, vegetation flourishes, completing a very brief life cycle.

In the centre of the northern continents, the climate is often harsh, with hot, dry summers and cold, dry winters. While they avoid the extremes of the desert or the arctic, they will support less complex communities than regions in similar latitudes which are closer to the oceans, where annual precipitation is generally higher and temperatures are moderated by the insulating effect of the sea. The original climax vegetation over much of north-western Europe, including Britain, was mixed, deciduous forest, with a high degree of species diversity. Man is a species found naturally only in the more complex communities, and we think of the climate that supports them as 'gentle'.

The richest communities of all are found in the rain forests

Sand retains water poorly and, close to the sea, blown sand such as this is made even more arid by the salt water. Plants such as marram grass (in the picture) have reduced their transpiration to a minimum. Their roots help consolidate the dunes, reducing further blowing and erosion.

of the wet tropics, where temperature and rainfall stimulate a very high level of plant productivity.

Farmers exploit the nutrient stored by a successional community. By clearing away the climax community in order to sow their crops, they reduce the number of species leaving niches to be filled. The plant and animal arrivals are regarded as weeds and pests and provided the farmer can keep their combined biomass low, the nutrient energy they would have made use of will be available to his crop.

Where land has been cleared of all plants and then left untended we can observe the successional process. All such land will be recolonized in time. Slag heaps and other industrial spoil tips will sustain plants, and even quarry workings that may offer little but bare rock and rubble will become overgrown. You need not search far to find wild land which its configuration reveals that once it was a quarry.

The study of gravel pits can be especially rewarding. Often they flood with water to form a pool surrounded by wild land, and both land and pool are colonized so that the successions of two distinct kinds of community may be observed, together with that of a third community, where land and water meet. No two communities are exactly alike but you may expect to see first a few wild flowers, such as the greater willowherb. Some short grasses may appear, followed by coarser grasses and then shrubs, broom, or gorse. Eventually there may be one or two trees. Rainwater draining into the pool will carry nutrients to stimulate algal growth which may give way to more complex aquatic plants. In time you may see frogs, newts, birds, small mammals, and in warm climates, reptiles. Untended ponds eventually become overgrown and dry out into marsh.

Such pools as this, resulting from mineral extraction, are very valuable habitat areas. They compensate to some extent for the many farm and village ponds that have become redundant and been filled in, cattle today being watered from mains supplies, and horses little used. For the pond to be stable, of course, the water draining into it must be nearly balanced by the water lost by evaporation and transpiration. A pond can never be very stable, however, owing to the sudden, violent effects of the weather which may dry it out, greatly increase its volume, or deoxygenate it.

ECOSYSTEMS

So far we have considered groups of organisms as communities. As is proper to an ecological approach, we have been concerned less with the physiology of individuals than with their behaviour as groups and with the interaction of one group with another. We have discovered the existence of definite hierarchical structures within communities and we have found that changes at any level are likely to cause reverberations throughout the whole. In examining pest and predator relationships, we found that an increase in biomass at one level will cause an increase at a higher level and that in most cases the situation will return to its previous state by a series of checks and balances that may be highly intricate. We have examined the flow of energy and nutrient through the community.

All these characteristics of living communities are shared with systems in general. A 'system' is defined scientifically, as *an organized or complex whole; an assemblage or combination of things or parts forming a complex or unitary whole*. A living, or ecological, system is called an **ecosystem** and is defined as *a system that is open for at least one property, in which at least one of the entities is classed as living*.

The study of the theory of general systems has led to the development of a 'systems approach' that is used to solve problems relating to subjects from the behaviour of cells to electronic communications to the planning of large cities.

Within a system, checks and balances such as we have described are known as 'feedback loops'. They can be represented by lines connecting organisms or events, and normally they will have a negative effect. An increase in food supply leads to an increase in population and the growth of the food supply is checked, and this in turn checks the growth of population. It is possible to represent diagramatically complex systems composed of actual physical quantities of materials which flow, sometimes with time delays, between points in the system, causal relationships that govern the direction of flow and factors that govern rates of accumulation, outside sources from which materials or energy come and 'sinks' to which they go. Provided the quantities and relationships can be numerated, it becomes possible to translate the diagram into a predictive mathematical model.

An ecosystem is any area that is reasonably well defined and which contains living organisms. It is open in the sense that it relates to other ecosystems from which organisms and nutrients may enter and to which they may leave. Taken together, all the ecosystems on the planet comprise the **biosphere**, which is itself a system and is sometimes called the **global ecosystem**.

It has been found that all systems have certain shared characteristics, which means that they may be studied by a similar methodology. The technique is to proceed by a series of logical steps based on observations made in the real world to the construction of a mathematical model, or simulation, of the system. When a faithful model has been constructed, its

An ecosystem is three-dimensional. As well as extending laterally over a defined area, it extends vertically, from the underlying rocks, through the subsoil and topsoil, to the communities of plants and animals living on the surface and in the air above it.

ocelot
(Felis pardalis)

behaviour can be observed under any conditions the modeller may seek to impose.

An ecosystem may be large or small, simple or complex. In practice, the size of a system that can be studied will be limited by the intellectual capacity of the ecologist or systems analyst, or, alternatively, by the storage capacity of the computer at their disposal, because ecosystem studies of large systems are often computerized.

The whole of the Sahara Desert may be regarded as a single ecosystem, or the entire rainforest region of the Amazon Basin, but you should not expect to see the results of detailed studies of them. They are far too large for ecological study, and although the desert supports fewer species than the rainforest, and is, therefore, somewhat simpler, it is still far too complex to be examined as a totality. Nor is it necessary to do so, for a number of representative samples will serve as well, and their behaviour will describe the behaviour of the larger systems of which they are typical parts.

There are obvious dangers in extrapolating from a part to the whole. A study of your little finger would provide inadequate data for predicting the whole body. The selection of areas that are genuinely typical is of prime importance. The size of these areas will depend on the system to be studied, and the detail required.

Ecosystems may be very small. A single drop of rainwater, held in a leaf, supports certain micro-organisms and it is an ecosystem. A small patch of your own skin is a rather complex ecosystem, and human skin has been studied as such. None of its inhabitants is visible to the naked eye, but you remove whole colonies each time you wash. There are rod-shaped bacteria, between 2 and 4 microns (μ) in length (a micron is one-millionth of a metre) and spherical bacteria, the cocci, which

We may study this forest ecosystem either as a whole or by drawing a line around an area of it and studying the survey area, extrapolating our results to describe the whole. Within the larger system there are many smaller ones. Human skin, or a drop of rainwater, are ecosystems.
Top inset. Human skin: 1 *Demodex folliculorum*; 2 coccus; 3 rod bacterium. *Bottom inset.* Organisms in water: 1 ciliate protozoon; 2 flagellates; 3 green alga; 4 chlorophyta.

are about 1 μ in diameter. Most adults are hosts to *Demodex folliculorum*, a microscopic mite with a long, worm-like body and four pairs of short legs, which lives in hair follicles around the nose and chin.

The differences between the living communities that compose equatorial rainforest and those found in tropical desert are immediately evident. Despite the fact that the two systems occur in roughly similar latitudes, the differences are caused by climate. The annual rainfall in the Amazon region is more than seven times greater than that in the Sahara. Similarly, the communities found in the arctic are very different from either the rainforest or desert. Again, the difference can be accounted for by climate. The summer temperatures in the Amazon region are more than double those in the far north of Canada or Siberia.

Thus, the world may be divided into broad climatic zones, each of which supports communities of plants and animals typical of that climate. These typical ecosystems are called **biomes**.

Europe supports several biomes. In the far north, adjacent to the polar snow line, there is tundra. To its south runs a belt of coniferous forest stretching from Sweden to the Pacific and from the Arctic Circle to between 50 and 60 degrees North. South of that again, grassland covers much of the area of central Europe and Asia, with considerable areas of mixed deciduous forest, especially in eastern France, Germany, Austria, and south-eastern Europe. Below the forests comes the Mediterranean region. Over most of lowland Britain the natural climax vegetational system is one of mixed deciduous forest.

Within a biome, communities will be generally similar, but the individual species of which they are composed may vary. Particular environmental circumstances will produce a general pattern, such as temperate or tropical forest or grassland, but composed of local species.

Four typical landscapes from: *Above* the tropics; *Above centre* temperate zone; *Below centre* the arctic; *Below* desert. The tundra, coniferous forest, and subtropical grasslands, or savannah, are other biomes which extend over large areas. Here, even the arctic landscape supports life in the sea and even on land, during the brief summer.

grey seal
(Halichoerus grypus)

shrimp
(Palaemon elegans)

periwinkle
(Littorina littorea)

barnacle
(Balanus balanus)

sentinel shell
(Assiminea grayana)

common whelk
(Buccinum undatum)

crucian carp
(Carassius carassius)

Within the main biomes, local climatic changes result in smaller biomes. These are found in mountainous areas and where human activity has induced local variations.

All life began in water; oceans, lakes, ponds, and rivers support their own ecosystems, each with its hierarchy of autotrophs, producers, consumers, and decomposers. There are differences, however. The reduced gravity means that unicellular plants are able to float close to the surface where they make best use of sunlight, and there is little need for large plants. It is these tiny plants which, together with the small organisms associated with them and feeding on them, form the plankton, that provide most of the photosynthetic production, and that are grazed by the herbivores.

Consumers, too, are able to move in three dimensions more freely than do land animals, and in ecological studies of aquatic systems, species tend to be classified according to the level at which they feed. There are bottom feeders and burrowers, organisms that float and those that swim, those that live among stones or plants, and the surface dwellers. Organisms become specialized for life at particular levels, the environmental variations being much greater than those on land between, say, the forest floor and canopy.

In deep lakes and the oceans, pressure at the bottom is much greater than at the top and no sunlight may penetrate. The ocean floor is a cold place, and it is consistently cold, remaining unaffected by weather and with only minor seasonal variations in temperature. Creatures adapted to such conditions can no more survive close to the surface than can surface creatures at great depths. Yet an ecological relationship exists between the upper and lower layers as organic detritus sinks to provide nutrient at the bottom and as dissolved nutrient salts permeate from the floor.

Freshwater is less homogeneous than the seas. In particular it is subject to wide variation in free oxygen content, muddiness, nutrient content, and pH (acidity or alkalinity). Fast-

The seal is a carnivorous mammal that travels far out to sea but breeds in colonies ashore. The shrimp buries itself in the sand in shallow waters. The shellfish, crustaceans (barnacles) or gasteropods, live on the shoreline. The crucian carp lives in stagnant or slow-moving waters.

moving rivers, slow-moving rivers, deep lakes, and shallow ponds all support different kinds of communities.

While the composition of the communities that go to form the major biomes is dictated by climate and soils, there is room for wide variation locally, in both land-based and aquatic systems.

The Peru Current brings cold water from the antarctic to the western shores of Chile and Peru before sweeping out into the South Pacific, where it is warmed and becomes the South Equatorial Current. The Peru Current has numerous upwellings which bring nutrients close to the surface and carry free oxygen down, and it supports a population richer and more complex than that found in the ocean to its west. Until recently it supported the anchoveta, a small fish used to make fishmeal for feeding farm livestock. Anchoveta catches reached twelve million tonnes (t) in 1968, one fifth of the total world fish harvest in that year. Cormorants also feed on the fish and their

Even in the most remote areas, oases usually have some kind of permanent human settlement. Much of the Sahara is not sandy but rocky, as is most of the American desert. When it rains, the desert is blanketed with vegetation which disappears as rapidly as it came.

droppings form guano, an important fertilizer for man who can collect it from the rocks, and for the ecosystem itself, because much of this recycled nutrient falls back into the sea. The anchoveta industry failed early in the 1970s, possibly because of overfishing and possibly because of a shift in the current. Cold currents are much richer ecologically than warm currents, which hold less free oxygen. The anchoveta may be returning.

The Peru Current is rather like an oasis in a desert. There is water deep beneath much of the dry Sahara. In some places underground aquifers encounter impermeable rocks and water is forced upward to create a natural spring. The additional moisture makes plant life possible and the plant cover encourages further growth by reducing evaporation losses. The natural oases have been extended by the human inhabitants of the desert. Shafts were sunk where the ground water came closest to the surface and water collected at the bottoms to form wells. Then tunnels were made linking the bottoms of the wells so that water drained toward areas to be irrigated. These extended oases are called *foggara* and legend has it that they were first made by Jews and Berbers long before the days of Christ. The ground water may have been located by observation of the habits of the desert locust, which lays its eggs in soil that is imperceptibly moister than the surrounding soil. Today the oases support many wild plants and some agriculture. Often they are the sites of towns.

It is not necessary to have a desert before there can be an oasis, only a variation in local climate or soil. Isolated valleys in the middle of high mountain ranges support ecosystems very different from those in higher, more exposed environments. In less dramatic form, this difference can be observed in Britain, and it is demonstrated clearly wherever agriculture occupies a valley bottom but the hillsides and tops are used only for rough grazing. Even on fertile lowland farms, isolated copses and small stands of trees, which may be natural in origin or planted by the farmer as windbreaks, are ecologically oases. In each case the ecosystem supported by the oasis is richer in species, and its total biomass is higher, area for area, than that in the surrounding environment. Ecologically, an oasis is small, rich, and isolated.

Clearly, the definition of a biome is arbitrary and is useful only in delineating broad areas with similar characteristics. The definition of an ecosystem is no less arbitrary. Areas of woodland are often chosen for ecological study, and a wood standing alone amid grassland is an ecosystem. Yet within the wood each tree is an ecosystem in itself. When much of Britain was forested, many of the forests in lowland areas were dominated by the common oak, *Quercus robur*. On more acid or rocky soils its place was taken by the sessile oak, *Q. petraea*.

The oak provides food and shelter for a wide variety of small plants, insects, mammals, and birds, each species being linked in a food web. Lichens are usually visible as evidence of bacterial colonization on the bark, and there may be mosses as well. Large fungi are common, especially on the main stem. The bark shelters insects, which provide food for tree creepers and woodpeckers. Other birds may nest higher in the branches and if the tree is tall enough the topmost branches may accommodate a colony of rooks. Old oaks are often hollowed, the decomposing wood inside the hollow providing food for a range of saprophytic insects and bacteria (living off decaying organic matter), and the hollow itself sometimes housing an owl, part of its diet consisting of the mice and small mammals living among the roots. Even the shade the tree affords is beneficial to some organisms. Grasses and small plants may grow at its foot free from competition from taller shrubs.

It is said that an oak tree may take 300 years to reach maturity, may remain a mature tree for a further 300 years, and then take 300 years to die. Its great longevity may encourage it to development into an ecosystem of greater complexity than is found in association with most trees. Trees grown commercially are felled after as little as twenty years in the case of some conifers. The redwoods, *Sequoia gigantea*, may live for more than 3000 years.

The ecology of a mature oak tree makes a fascinating study, but it illustrates the open-endedness of much ecological fieldwork. Does the grey squirrel that moves from tree to tree or the fox and the badger that live in the woodland habitat form part of the ecosystem? **Key:** 1 oak bark beetle (*Scolytus intricatus*); 2 goat moth (*Cossus cossus*); 3 nut weevil (*Curculio venosus*). 4 great spotted woodpecker (*Dendrocopos major*); 5 oak gall wasp (*Biorhiza pallida*).

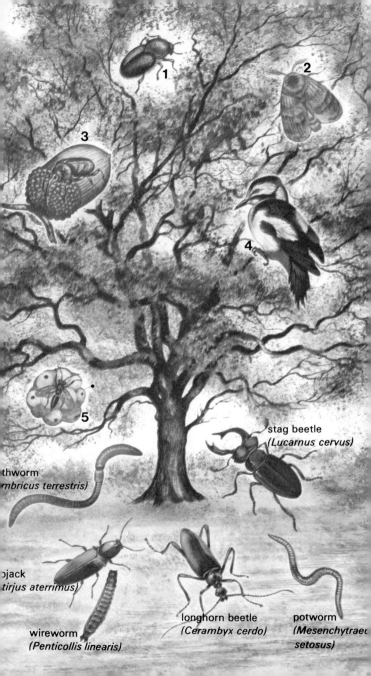

1

2

3

4

5

stag beetle
(Lucarnus cervus)

thworm
mbricus terrestris)

ojack
tirjus aterrimus)

wireworm
(Penticollis linearis)

longhorn beetle
(Cerambyx cerdo)

potworm
*(Mesenchytraeu
setosus)*

HABITAT AND ADAPTATION

The study of food webs and plant and animal communities leads naturally to the study of ecosystems and the dynamism of the relationships within them. Other approaches are possible, however. If we begin with a particular species and study its requirements for food and shelter we enter the ecosystem, as it were, from the other end. The total requirement for a species constitutes its **habitat**. The study of habitat is of particular value in conservation, because it is possible to manipulate an environment to make it more or less hospitable to species where we wish to increase or reduce populations.

We may use our knowledge of habitat to predict the presence of species. We would not expect to find a polar bear in a European forest, for example, because we know that polar bears require an arctic habitat, but we might expect to find species of European bears which inhabit forests.

We may look for the otter close to water. The fresh water otter lives in a hole in the bank of a river or pond. It swims well and its diet consists mainly of fish.

The badger is found at the edge of woodland. It lives underground, in a sett, and feeds at night on small rodents, slugs, insects, beech nuts and acorns and, occasionally, the egg or chick of a ground-nesting bird.

The fox is found all over Europe. It, too, lives underground and hunts small mammals at night, but it is a nomad, the male seldom staying for long in one place and the vixen doing so only while she is rearing her young.

Species may become extinct within a particular ecosystem for a variety of reasons, but if the habitat is suitable it is sometimes possible to re-introduce species. The most spectacular success in this field has been the re-introduction of the néné, or Hawaiian goose to Hawaii. Specimens of this threatened bird were bred at the Wildfowl Trust, Slimbridge, Gloucestershire, England, and exported.

The polar bear, the otter, the badger, and the fox are all

Historically, naturalists studied habitats long before ecologists began to construct theories to explain the relationships between species. In natural history museums it is still usual to find stuffed animals and birds shown in a reconstruction of their natural habitat.

rey squirrel
(*Sciurus*
arolinensis)

fox
(*Vulpes*
vulpes)

badger
(*Meles meles*)

otter
(*utra lutra*)

trout
(*Salmo trutta fario*)

mammals and share many features in common. They are all warm-blooded, viviparous (giving birth to live young) vertebrates and their internal organs are arranged in a rather similar fashion. Yet they are very different in appearance and habit. Some mammals are herbivorous, some omnivorous, some carnivorous, some large, some small. Each species has adapted to a particular kind of environment. The existence within an ecosystem of a supply of food or shelter constitutes a niche, and when a species exploits it, that species fulfils an ecological rôle within the system. It consumes food, cycles nutrient, provides food for others.

No two systems are exactly alike, and similar niches may be filled in a variety of different ways. In the same way species may adapt to fill niches. The thirteen different species of finch that Darwin found in the Galápagos Islands all descended from a single species. Six species had adapted themselves to life as vegetarians on the ground. Five more became insectivorous tree finches and another became a 'woodpecker'. Each species found a niche or made one.

Similar niches in different ecosystems may be occupied by organisms that bear no relationship to one another but which, because of the rôle they fill, come to resemble one another in behaviour and appearance. The whale resembles a fish, for example, but it is a mammal, and the seal is related to the cat. This is called **convergence**.

Moorland, found in northern and western Britain at altitudes between 305 m and 610 m, provides a habitat for only small plants. The soils are acid and often peaty, with the peat very deep in some places. Biological productivity is low, the plant community being dominated by coarse grasses, heather, and bracken, with some shrubs such as the crowberry, cowberry, and bilberry. There is a wide variety of bird species, ranging from those that nest on the ground to the hawks. Some moors support wild horses.

Very little remains of the forest that once covered Britain.

This landscape is typical of much of the Highlands of Scotland, and these are some of the species you may hope to find there. Reindeer are being reintroduced in places, Highland cattle are still half wild, wild cats still survive, and the birds are dominated by the golden eagle.

buzzard
(Buteo buteo)

skylark
(Alauda arvensis)

bracken
(Pteridium aquilinium)

red deer
(Cervus elaphus)

red grouse
(Lagopus
lagopus
scoticus)

heather
(Calluna vulgaris)

lichens

bank vole
(Clethrionomys
rutilus)

moss

It was cleared for agriculture, but where land has reverted to wilderness, the natural succession has tended to be towards a forest climax. Consequently, although the forest has disappeared, ecosystems somewhat similar to those found in a forest are common. The trees and shrubs planted to form hedgerows and shelter belts resemble those found at the edge of woodland and apart from routine maintenance, rural hedges are left undisturbed, especially at their centre, for years or sometimes centuries. Some of the hedges on British farms were planted in Tudor times and the majority were planted in the 1700s and early 1800s. The age of a hedge can be estimated from the number of plant species found in a measured length.

The interface where two ecosystems meet, called the **ecotone** supports species from both systems and permits migrations, and it is often richer in species than the centre of the main systems. Woodland edge is richer than woodland itself. Almost all the main groups of British wildlife are found in hedgerows and shelter belts. Nearly 1000 native plant species have been recorded there, most of the resident woodland birds will breed in hedges, and many others feed in them. Nearly half the country's resident mammal species may be found in hedgerows as well as more than twenty species of butterfly and many other insects. In a typical hedge you might find oak, birch, elm, alder, aspen, willow, pine, or wild cherry trees; hawthorn, blackthorn, hornbeam, rowan, holly or crab apple shrubs; and such flowers as bluebells, dog's mercury, and primroses.

Today, many British farmers manage their hedgerows so as to make them the best habitat possible. They may be laid at

Similar, ecologically, to woodland edge, the hedgerow is the richest habitat in lowland Britain. Apart from its rich flora, the hedgerow provides a pathway linking small woodland areas, preventing their isolation. The widespread clearance of hedges is an ecological catastrophe.

Key: 1 song thrush (*Turdus philomelos*); 2 hazel hedge (*Corylus avellana*); 3 beech (*Fagus sylvatica*); 4 elder (*Sambucus nigra*); 5 stoat (*Mustela erminea*); 6 bramble (*Rubus fruticosus*); 7 nettle (*Urtica dioica*); 8 rabbit (*Oryctolagus cuniculus*); 9 dandelion (*Taraxacum officinale*); 10 hedgehog (*Erinaceus europaeus*); 11 sow thistle (*Sonchus oleraceus*); 12 bracken (*Pteridium aquilinium*); 13 common shrew (*Sorex araneus*).

the bottom, to make them stock-proof, but they are trained and trimmed into an inverted V shape about 2·5 m high. A strip of grasses and herbs is allowed to grow wild on either side, so that cover is provided for small mammals at the bottom, nesting for birds higher up, amid many plants.

Provided the supply of water and nutrient is adequate, the more energy that flows into an ecosystem, the more efficiently the nutrients will be utilized. The more efficiently nutrients are taken up by plants, the greater will be their biological productivity, the more diverse the plant cover, and the greater the variety of consumer species.

Over much of the northern part of South America, parts of Central America, west-central Africa and parts of south-east Asia, including Indonesia and Papua, annual rainfall exceeds 1900 mm and there is less than 6 °C difference between summer and winter temperatures, around a mean of 30 °C. The climax vegetation is tropical rainforest, and its production has been estimated as 32·5 t/ha/yr. Oak forest production, measured by the same method, is 9 t/ha/yr.

The ecosystem is dominated by tall trees, which shade smaller plants. While this would inhibit growth in a temperate climate, in the tropics there is still enough light and warmth for a luxuriant surface cover. The forest canopy also shelters smaller plants from the physical impact of the rain, which is often intense.

There is a proportional diversity of animal consumers, although the large mammals such as the elephant, lion, giraffe, and rhinoceros, live on the open savannah. In African rainforest you may nevertheless find buffalo, pigmy hippopotamus, and in rivers and swamps, hippopotamus, several species of small antelope, wild pigs, spotted hyena, and *Canis adustus*, the side-striped jackal. It is also the home of the large constrictor snakes.

The tropical rainforest is also the habitat of many of the large primates, including the forest gorilla, the chimpanzee and, in Asia, the orang-utan. It is also the home of their distant relative, man, and primitive peoples still live as hunters and gatherers in the rainforests of Africa, South America, and parts of Asia, wherever they have avoided close contact with more advanced cultures.

Only a small part of the incoming solar radiation is used in photosynthesis, but plants will tend to maximize their exposure to light. In a climax ecosystem, the leaf canopy is complete as the foliage from one plant meets that of the next.

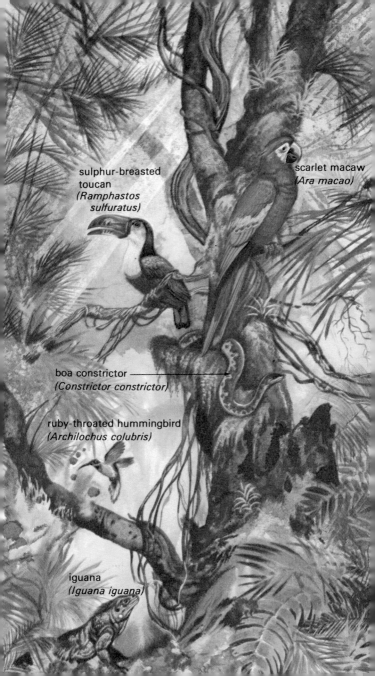

sulphur-breasted
toucan
*(Ramphastos
sulfuratus)*

scarlet macaw
(Ara macao)

boa constrictor
(Constrictor constrictor)

ruby-throated hummingbird
(Archilochus colubris)

iguana
(Iguana iguana)

Compared with a tropical rainforest, a mountainside seems a very poor, simple habitat. Nevertheless, it is far from lifeless. In Britain, land above 600 m supports heathland or grassland communities, and above them the mountain-tops sustain communities similar to tundra.

Mountain heaths are usually dominated by heather (*Calluna vulgaris*) or bilberry (*Vaccinium myrtillus*). They are less common than grasslands because the practice of grazing sheep over them tends to destroy the heather and allow it to be replaced by grasses. Mountain grasslands are widespread. Usually they are dominated by a few grass species, such as purple moor-grass (*Molinia caerulea*), sheep's fescue (*Festuca ovina*) mat grass (*Nardus stricta*), and bent (*Agrostis*). There are many herbs associated with the grasses and the remoteness of many of the habitats has tended to protect some of the rarer species. In Wales you may find moss campion (*Silene acaulis*) and purple saxifrage (*Saxifraga oppositifolia*). The tundra-like regions of the mountain tops are less stable, owing to the combined action of frost and erosion by wind and rain. You may find dwarf willow (*Salix herbacea*) there, and reindeer moss (*Cladonia rangiferina*).

Biological production is low. True tundra produces only 1 t/ha/yr. This level of productivity will support only small consumer populations, dominated by insects and birds that can travel widely in search of food. Close to the tree line, however, the ecotone provides a richer habitat, mainly due to migrations from the forest to the mountainside. The afforestation of the lower slopes improves the habitat at higher altitudes.

There is a theory that the yeti, apparently a large, unidentified primate occasionally encountered in the Himalayas, may spend most of its time in a deep, remote valley in a forest habitat, and that it may be a species previously believed to have been extinct for millions of years.

Temperature falls with increasing height, and a mountain range may reproduce main climate zones on a small scale. In a tropical range it is possible to proceed from the tropics, at sea-level, through subtropical, temperate, coniferous forest, tundra, and finally to arctic systems.

golden eagle
(Aquila chrysaëtos)

ibex
(Capra ibex)

coniferous forest

chamois
(Rupicapra rupicapra)

mountain hare (showing winter camouflage)
(Lepus timidus)

In the world as a whole, the most widespread climatic type is that associated with desert or semi-desert. It is not possible to state the minimum rainfall needed to prevent extreme arid conditions, because what is important is not rainfall alone, but the amount of water in the soil. This is determined by the rate of evaporation as well as by rainfall, so that a climate that will support extensive vegetation in high latitudes will not do so closer to the equator. In North America, about 500 mm of rain are needed to prevent extreme aridity in North Dakota, and about 750 mm in Texas. Less than 250 mm of rain a year will produce desert almost anywhere. The average annual rainfall in the Sahara is 100 mm, yet the Sahara is by no means the world's driest desert.

The rainfall occurs in short, intensive downpours which fill the river beds, cause extensive erosion that prevents the development of fertile soil, and which leave dissolved salts on the surface as evaporites. Xerophilous plants are those which prefer such conditions of drought and high salinity. They economize in their use of water either by developing leaves that are tubular in shape, with the stomata, through which water is lost by transpiration, on the inside, or as thorny bushes. Xerophilous grasses are found growing sparsely in steppe semi-arid conditions, and on sand dunes close to the coasts of Europe. The most common thorny bush of the Sahara is the tamarisk, *Tamarisk articulata*, and acacias are desert plants. A range of sparsely distributed animals consumes this vegetation, including many insects, reptiles including in America the rattlesnakes, and small rodents.

The arctic and antarctic are covered with ice and snow because they retain such precipitation as they receive, but these regions are very arid. The mean annual rainfall between 80 and 90 degrees North is about 100 mm. There has been no accurate sustained measurement at the same southern latitude, but between 70 and 80 degrees South, rainfall averages 76 mm.

The saguaro cactus may grow to a height of 15 metres. True cacti are found only in North and South America, although members of the genus *Rhipsalis* are found wild in the Old World and may be native in parts of Africa and Sri Lanka. The cacomistle is a carnivore, related to the raccoons.

saguaro cactus
(Cereus giganteus)

desert rabbit
(Sylvilagus floridanus)

cacomistle
(Bassariscus astutus)

desert fox
(Vulpes velox)

kangaroo rat
(Dipodomys)

CRITICAL ECOSYSTEMS

Just as there are areas of ecotone at the interface between two local ecosystems, so there are boundary zones separating the world's major biomes. Unlike local ecotones, however, these 'buffer' zones are large enough to be considered as major ecosystems in their own right.

Separating the arctic and antarctic icecaps, which are cold deserts, from the high latitude limits of the forest belts, there is tundra. It is a system characterized by an extensive cover of small, low plants, mosses, and lichens, growing on permafrost-

See also map on facing page. Areas that are particularly sensitive to ecological interference. Generally, these are the relatively small zones that form a boundary, and buffer, between major biomes and ecologically are quite different from them. Many of the species associated with them could not survive elsewhere; if the physical characteristics of the buffer zone were to alter, not only would the flora and fauna change, but much of it might disappear completely.

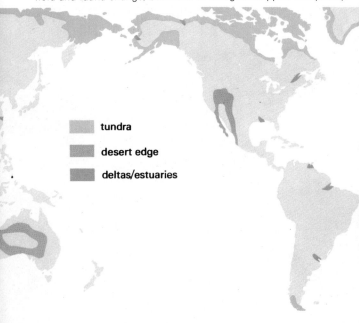

tundra

desert edge

deltas/estuaries

ground with a surface which may thaw in summer, allowing a concentrated burst of plant growth, but which is frozen below the surface and to a great depth. Tundra supports a number of animal species.

Large deserts are not uniformly dry and there are many local ecological variations. Bounding the generally arid zones, and separating them from the savannah-type grassland, or steppe, that is the neighbouring biome, there is a semi-arid area of poor vegetation, supporting a sparse kind of pastoralism. On the southern edge of the Sahara, this is called the Sahel region.

Estuaries and coastal wetlands form a boundary between land and marine biomes and between fresh and salt water. While tundra and desert edge occur next to systems with low productivity, wetlands border two or three systems that may be highly productive and derive nutrients and species from each. Like true ecotones, they are highly productive.

Tropical and subtropical forest bounds the rainforest, grading into the savannah. It is probable that this is the environment in which the ancestor of man first appeared, as a large-brained primate forced to abandon his arboreal habitat as the forests contracted during the climatic cooling of the late Miocene and Pliocene, ten to twenty million years ago.

These buffer zones are ecologically fragile. Comparatively small changes may lead to their absorption into one or other of the larger systems, or to their destruction.

Fears have been expressed, for example, that an extensive surface cover of dark-coloured material, such as oil, over the tundra could increase the amount of solar radiation absorbed during winter, when the ground is covered with snow. This could cause a general warming that might be sufficient to begin to melt the permafrost. Such a process might be self-perpetuating, as snow and ice were replaced by water and mud, leading to further energy absorption. The end result might be extensive waterlogging which would destroy surface vegetation and deprive most animals of their habitat. Alternatively, any slight climatic cooling could extend the icecaps, deprive the tundra of its short period of summer warmth, and absorb it into the arctic proper. In this case the tundra might re-establish itself at a lower latitude as the edge of the forest area

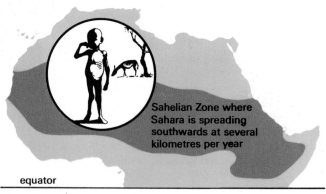

Sahelian Zone where Sahara is spreading southwards at several kilometres per year

equator

The climate change in this area will probably last for at least a century, and the climate of the whole northern hemisphere has been affected. In Britain, rainfall has decreased and anticyclonic weather has become more common. The Sahel Zone may become uninhabitable to humans.

was pushed back. A waterlogged tundra would develop its own, different, ecosystem in time.

Such a climate change appears to be happening at the present time, affecting most dramatically the desert edge from Central America, across the Sahara, the Near East, and parts of the Indian subcontinent. A slight cooling in the northern part of the northern hemisphere has reduced the amount of solar energy available to power the movement of large air masses. So weather systems have weakened. There has been substantial cooling at the poles, and the circumpolar vortex, a continuous westerly flow of winds in the upper atmosphere, has moved south. This has pushed the polar front to the south and the subtropical atmospheric high-pressure areas moved closer to the equator. It is these highs that bring dry air down from high altitudes and cause arid conditions, and it is their southern edge that marks the northern limit of the monsoon rains. Therefore, the monsoon belt has shifted to the south. The semi-arid desert edge has experienced a sharp decrease in its annual rainfall, turning it into a true desert. Since about 1960, the Sahara has been advancing to the south about 9 km a year and it is likely that similar movements are occurring along the edges of all the northern hot deserts.

These regions are inhabited by nomadic pastoralists. The failure of the vegetation has caused livestock to starve and led to widespread famine. Similar losses have been sustained by wild animals.

The same process has produced slight changes in climate in Europe. As the north polar icecap has been observed to be growing, and as Alpine glaciers have been advancing, mean temperatures in Europe have fallen. The British Isles are covered by a tongue stretching south from the arctic anticyclone and this has led to a marked reduction in westerly weather and an increase in anticyclonic weather. Rainfall has been diminishing over the country as a whole but especially in eastern England and Scotland, where drought has interfered with crop growing. In general the weather has been characterized by prolonged dry periods, broken by violent rains. If the trend continues, the growing season may grow a little shorter and the climate rather drier and cooler. Probably this is a natural change, after a period of warming from the mid-1800s to the 1940s.

Many species of marine fish spend the early part of their lives close inshore and in the mouths of estuaries. Thus, estuaries play a vital rôle in the ecology of the sea. A true estuary is formed when a river flows down a shallow gradient into the sea. The flow is weak, and the tidal movement of sea water is able to overcome it, bringing salt water into the river, often to a considerable distance upstream; the river flow resumes as the tide ebbs.

Salt and fresh water mingle only slowly. If the tidal pressure is strong, sea water may enter in a compact mass. There will be some mixing of salt and fresh water at a front. If the flow is weaker, fresh water will tend to ride above the denser sea water so creating a vertical salinity gradient, and salt water species will be able to travel upstream beneath the fresh water.

Rivers carry silt, some of which is deposited to form mudbanks. Where the two waters meet, the turbulence makes the water muddy. Light penetrates poorly and photosynthesis is inhibited. The temperature of river water is subject to much wider variation than that of sea water. Thus, the estuarine environment is one of extremes, but it is also rich in nutrients, and euryhaline animals, which can control the concentration of salts within their bodies, have ample food. They are mainly amphipods, such as *Gammarus* and *Corophium*, prawns such as *Palaemonetes varians* and *Praunus flexuosus*, ragworms such as *Nereis diversicolor*, and snails such as *Hydrobia ulvae*. Estuaries also support both green seaweeds, especially *Enteromorpha*, and brown seaweeds such as *Fucus vesiculosus*, *F. spiralis*, *F. serratus*, *Pelvetia canaliculata*, and *Ascophyllum nodosum*.

Apart from the animals adapted to life in brackish water and on mudflats, there is a gradual change from fresh- to sea-water species and some species can survive in either environment. The flounder and *Lophius piscatorius*, the monk- or angler-fish, can pass through estuaries and live in rivers. Eels, salmon, sea-trout, and shad use their tolerance for either habitat to spawn in one and spend most of their lives in the other. The salmon, sea-trout, and shad live in the sea but spawn in rivers, and the eel lives in rivers but spawns at sea.

Salt marshes, which are often found in estuaries, support the growth of rough grasses and some flowering plants and they have their own populations of adapted insects and crustaceans as well as those of estuaries. They provide important feeding grounds for certain birds.

The shore-line itself supports many species of plant and animal, although it, too, is a harsh environment. On muddy coastlines in the tropics, as well as on mudflats, mangrove trees (*Rhizophora mangle*) may become established. They spread through the shallow salt water by sending out adventitious roots, which descend in an arch from the parent stem and send up new stems from the point at which they strike ground. The seeds send down long roots before the fruit is detached from the tree, so that when the seed falls, it is already rooted. In this way very complex root systems develop which afford shelter to a

A number of important sea fishes spend a part of their lives close inshore, so that estuaries and coastal wetlands have an ecological importance disproportionate to their area. The sea lamprey is a parasite, having no jaws but a round, toothy sucker with which it attaches itself to its host. It wounds the host and ingests its blood.

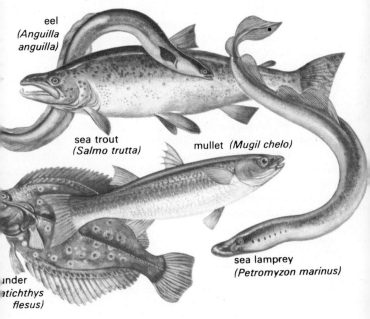

eel
(Anguilla anguilla)

sea trout
(Salmo trutta)

mullet (Mugil chelo)

sea lamprey
(Petromyzon marinus)

flounder
(Platichthys flesus)

wide range of species, so encouraging the development of rich ecosystems. Mangrove swamps are found off the Florida coast and in India.

The ecology of estuaries depends on the quality and quantity of water they receive, and changes in river flow or composition may produce large effects.

Even in the major biomes, it is possible for changes to initiate self-perpetuating processes of deterioration. When rainforest is cleared, for example, the surface is no longer sheltered from the sun and rain. Soil temperatures rise, causing rapid drying out during dry periods, and the physical impact of the rain increases, leading to soil erosion. There is little storage of nutrient in tropical forest soils and such nutrients as there are leach out. It is difficult for other vegetation to become established. Transpiration is reduced and if the deforestation extends over a wide enough area there may be local climate changes, with the development of arid desert.

Many tropical soils are lacteritic. Laterite (from the Latin, *later*, a brick) is composed of oxides and hydroxides of iron and aluminium forming layers or nodules in the soil. These layers are impermeable and water accumulates above them. When vegetation is cleared, repeated waterlogging and drying out accelerate erosion and plant nutrients are leached away. When ferric oxide is dried it hardens irreversibly into a rock-like substance, while the aluminium compounds form part of soft clays beneath the laterite. No further plant growth is possible.

Parts of Mexico once supported rainforest which was destroyed by repeated burning and became arid desert. Lateritic soils are distributed widely in tropical South America, west and central Africa, parts of south-east Asia, and northern Australia.

There are no true lateritic soils in Europe, but the removal of forest has led to near-desert conditions in many areas that are

Above Forest fires occur most commonly in subtropical and temperate zones, where the climate is more likely to permit the drying out of the vegetation. *Below* Repeated burning that allows insufficient time for the forest to re-establish itself, always causes severe erosion, and may create conditions of extreme aridity and acidity, even where the soil is not lateritic.

fossils

1

2

3

4

fossils

1

2

5

3

4

now heath or moorland. In Britain, the north Yorkshire moors and Dartmoor were once forested.

It is possible to reconstruct the vegetation patterns of the past. The work is performed by botanists collaborating with archaeologists. Archaeologists are able to date accurately the levels at which they are working. Botanists take soil samples from those levels and extract from them grains of pollen, which are preserved in the acid soils for thousands of years. By identifying the plants from which the pollen came it is possible to build up a picture of the species present at that time, and of their distribution.

In general, until about 4000 years ago, our heaths were forests dominated by oak, birch, and alder, with some hazel. Bronze Age farmers made clearings by burning, and their animals grazed the open spaces, sometimes preventing the return of the forest. When trees did come back there was more hazel and birch and less alder. Then, with further clearing, the trees disappeared altogether, to be replaced by hazel bushes, grasses, and bracken. There was some intermittent cereal cultivation at this time. The organic matter was lost from the topsoil and nutrients leached out quickly. The soils became ash-coloured sand. Man-made fires were accompanied by natural fires among the grass and bracken. About 500 BC the climate became wetter, favouring heather and accelerating the leaching of nutrients, leaving the acid, barren soils we have today, suitable for afforestation, but not for agriculture.

Some long-living trees, such as the Douglas fir (*Pseudotsuga taxifolia* or *menziesii*), some junipers and pines, are sensitive to changes in temperature and rainfall, which affect their rates of growth. By a study of their annual rings it is possible to relate the amount of growth to climatic conditions, coupled with ring-dating or dendrochronology, to reconstruct past climates.

Mixed deciduous woodland once covered most of Britain. The vegetation patterns of the past can be reconstructed from pollen grains and other plant remains of a determinable age.
Most of the forest was cleared from about the 1500s. **Key:** *Above* mixed woodland 1 fossil fruit; 2 catkin; 3 seed - cinnamon; 4 pollen grain; *Below* open moorland 1 seed; 2 pollen grain; 3 seed - water soldier; 4 pollen grain; 5 seed.

MAN'S ENVIRONMENT

Many species modify their environment to make it more hospitable. Ants, termites, bees, and other social insects construct elaborate communal nests. Beavers build dams. Man, too, has learned to alter the natural environment so that it will support him in greater comfort and security.

Our far distant ancestors, the man-like primates, lived in the tropical forest. Probably they spent much of their time in the trees, where food was plentiful and enemies few. Later, they were forced to live in more open habitat, perhaps as a result of climate change that killed off much of the woodland. The new environment was hostile. Fruits were less plentiful and there were many large, skilful and determined predators. Ape-like man learned to eat meat and to hunt and fight in groups. His social structures became more complex. Eventually he experimented with 'taming' the hostile world, with making it a better place for him to live.

Thus far there is no qualitative difference between man's attitude to his environment and that of any other species, and his behaviour is that of an opportunist making the best of the conditions in which he finds himself. As time went on, however, a quantitative difference became apparent. It was not what he did that was different, but the scale on which he did it. He cleared the forest from vast areas and built great cities. He tried to regulate certain natural processes with his technology, such as irrigation. When he failed he created desert in lands that had once been fertile.

This, too, is part of a natural process because man, as a species living within the natural system, cannot behave 'unnaturally'. In evolutionary terms, it may be that as an opportunist species, never well adapted to life outside the forest, he has used his assets – a large brain, agile hands, and an ability to subsist on an omnivorous diet – to construct his own environment, more favourable to him than that into which nature had forced him.

This is another kind of succession. Trees are cleared to provide farm land and space for a village. The village becomes a town and the forest becomes the gardens and parks within it. As the value of urban land rises, building density increases at the expense of open spaces and trees.

There have been three steps of major significance in the development of man: the acquisition of speech, fire, and agriculture. Fire made it possible to clear the forest exposing fertile soil for easy cultivation and grazing, and without speech it would not have been possible to convey complex information about plants and animals and to increase the store of such information within the tribe.

Agriculture probably began in semi-arid regions, with many different small ecosystems and consequently a food supply that was diverse and generally plentiful, but that could suffer from seasonal drought. As a plant-gatherer, early man knew much about the plants he collected. His cultivation of them began in natural open spaces, such as stream terraces and dump heaps. In the Old World, the cereals wheat and barley were among the first plants to be domesticated. In the New World the Amerindians grew maize and cucurbits (gourds and melons). The domestication of plants began about 12 000 years ago.

The development of agriculture was a major step in the development of man, but ecologically it was often of only minor significance. The area of forest edge ecotone has been increased and so has the diversity of species that grow in the clearing.

The change to modern industrial farming is of less importance to man, but the ecological change is profound. The forest has gone, reduced to one tree growing beside a house, there are no hedgerows, and a very small range of plant species occurs over large areas.

In modern times, the industrialization of many crafts has been followed by the industrialization of agriculture, with extreme specialization and an increase in the intensity with which the land is used. In arable regions, livestock have been moved into indoor units and the trees and hedgerows that once sheltered them removed. Only plant crops, mainly cereals, are grown.

Early agriculture represented a major ecological change. Often the partial clearance of woodland increased the area of ecotone and so improved the habitat. Modern farming is another change of equal significance, but one that causes the deterioration of wild habitat.

Early farmers practised slash-and-burn, or swidden agri-

The medieval European city retained orchards and large vegetable gardens, with some livestock, within the walls. It was Ebenezer Howard (1850–1928) who reintroduced large areas of park and garden into his garden cities, starting with Letchworth and Welwyn in Hertfordshire, England.

culture, probably combined with the hunting of wild animals for meat, bones, and hides. In many societies, men hunted while women grew crops. A settled existence became possible with crops being grown on fields around the village. Surpluses of food were produced, which allowed for the support of individuals not engaged directly in food production or gathering, and so occupations became more specialized.

As agriculture spread and the population grew, villages became small towns. The specialization of occupations encouraged trade and the towns became trading centres and communications centres, linked by networks of tracks and roads. Often the town occupied a position that could be defended against attack from neighbouring tribes and fortified towns provided security. From being trading and communications centres, towns became centres of political power, and tribal chieftains became kings, dukes, and princes.

The development of settled agriculture led to the development of cities and city states, and made civilization possible. At its height, the agricultural system of the Fertile Crescent in Mesopotamia supported urban populations whose densities were similar to those found in Europe today. Yet the 'hydrological' civilizations, depending on sophisticated irrigation technology in basically arid climates, generally failed, either through waterlogging of the soil due to inadequate drainage, or to salination due to the accumulation of salts left at the surface by the evaporation of water rising from the water-table. Both the Babylonian and Indus Valley civilizations failed for this reason.

European cities prospered, however. Our word 'urbane', meaning courteous, suave, or elegant, comes from the Latin *urbanus*, of the city, and our 'civilization' comes from *civis*, citizen. Eventually the agricultural systems of the Greek city-states and even of Rome itself, came close to failure owing to the over-exploitation of the land because food was consumed far from its site of production, and biological nutrient cycles were broken. It is said that the fertility of Sicily was lost in the sewers of Rome, and over parts of the northern Sahara the outlines of Roman fields can be seen to this day. Nevertheless, the cities have given us our works of art, our philosophy, and our science, and we belong to an urban culture.

Since urbanization first began, the cities have attracted people from rural areas. If man first appeared about three million years ago, and agriculture began about 10 000 years ago, then man's experience of farming amounts to no more than 0·33 per cent of his time on Earth, and his experience of urban life to no more than 0·166 per cent. Yet in this brief interval he has moved from Jericho, one of the oldest cities but one which affected the lives of very few people, to the present situation, in which we are fast approaching a point where half the world's population lives in cities. The very speed with which this has taken place may suggest that city life is irresistibly attractive, and that our future is to be city-dwellers.

It was not the farmers who completed the clearance of most of Europe's great forests, however, but the industrialists, who needed timber for shipbuilding and fuel.

Man learned to extract and work metals at the beginning of the Bronze Age, about 4000 years ago. It is believed that tin was extracted from cassiterite (tin oxide) in Cornwall and Devon around 1800 BC, and mining prospered and expanded during the Roman occupation of Britain. The south-west of England possessed rich deposits of many metals, but little fuel for smelting, and the energy required to refine the elements from their ores was provided elsewhere. The forests were

As the richest mineral ores are used, it becomes necessary to extract from lower grades, pulverizing very large quantities of rock. This is a copper strip-mine. The vegetation and topsoil are removed from several square kilometres and may not return for centuries after mining has finished.

being cleared and their timber used faster than it could be regrown. It was obvious that alternative sources of energy were needed. Coal was being burned in London in medieval times and during the 1700s and 1800s it became the most important fuel.

As surface deposits and higher grades of mineral ores were used up, it became necessary to mine beneath the surface and to use lower ore grades. This called for more work; therefore, more energy and more fuel were needed. The industrial nations became dependent on fossil fuels.

The fossil fuels – coal, oil, and to a lesser extent, peat – represent solar energy that was 'trapped', or fossilized, when geological and climatic conditions prevented the complete oxidation of organic matter. When heated, they burn rapidly, completing the oxidation process and releasing heat that came originally from the sun. In modern times, man has begun to derive some of the energy his industries need from uranium – a metal that is extracted from low-grade ores – and from geo-thermal energy – the heat of the Earth's own core.

By his use of metals, man modified his immediate environment. New, more efficient tools helped him extend and improve his agriculture, building, and communications. At the same time they facilitated further modifications in the future. The metals and fuels derived from high-grade, easily mined deposits gave him the capacity to work lower grade ores and to extract fuels from greater depths. Small market towns have grown into vast industrial conurbations, tracks through the forest have become motorways and railways running across arable land. Today, it is feasible to extract oil and gas from deep beneath the sea bed and the most highly priced metals are extracted from very low-grade ores by stripping away surface vegetation and soil and pulverizing rock, cubic kilometres at a time.

Oil, coal, and natural gas are not renewable. The world possesses very large reserves of oil and coal, but rising prices and the fear of shortages have stimulated intensive research into alternative sources of primary energy. Direct solar heat may be used for large-scale projects and certainly it can be used for domestic purposes. Wind and tidal power are also being considered. In each case it is necessary to evaluate the 'environmental impact' of innovative technology. These natural energy sources involve no combustion of materials so that they are likely to be less polluting, although tidal barrages may affect estuarine ecosystems.

All of man's deliberate modifications of his environment have been designed to improve his standard of living. Modern man lives far more securely than did his ancestors. To a large extent the threat of famine has been removed, at least for the peoples of industrial nations, infectious disease has been controlled and the average expectation of life at birth for Europeans has doubled since the Middle Ages. Improved communications and transport have enriched life by enabling us to experience at first hand climates, environments, and lifestyles our grandparents could know only from books.

Yet there has been a price to pay for these improvements. By making the environment more hospitable to us, we have made it less hospitable to other species. Our gain has been

Rivers may be damaged by poisoning or by deoxygenation. Many industrial effluents are toxic, but warm water reduces the amount of free oxygen in the river, as does the decomposition of organic matter, such as sewage. Most industrial nations now try to improve the condition of their rivers.

nature's loss. Plants and animals that cannot adapt to environmental changes die out and the rate at which species become extinct has increased rapidly since the beginning of the Industrial Revolution.

There have been other environmental changes brought about as by-products of man's activities. The combustion of fossil fuels at an ever-accelerating pace has caused local heating of air and water and has released particles and gases into the atmosphere. Particles may blanket out some incoming solar radiation; the gases include sulphur dioxide which dissolves in clouds and falls as acid rain to cause subtle environmental changes, and carbon dioxide, which may trap radiation leaving the planet's surface and which increases rates of photosynthesis. Toxic effluents from industry have changed aquatic ecosystems and soils have been eroded by over-intensive agriculture.

The sea is also affected by sewage, but the discharge of oil by accident or from the cleaning out of tanks is widespread and very harmful to marine life, possibly reducing photosynthesis. Oiled seabirds can be cleaned and released, but only a few are saved in this way.

PROTECTING THE ENVIRONMENT

As our modification of the global environment continues, it becomes increasingly important for us to understand its ecosystems and our place within them. The more thorough our knowledge of human ecology, the less likely we are to initiate trends that will render our environment more, rather than less hostile.

We must begin by studying ecosystems that are not subjected to human interference, in the remote, wild places where relationships between species establish themselves, and are maintained, by the interaction of solar energy and water with the internal dynamism of the systems.

The increase in man's numbers, his mobility and his industrial activity lead to constant human expansion and a consequent reduction in wilderness areas. In Europe there is very little true wilderness left, although areas once occupied by man but later abandoned may also be of scientific interest, as may ecosystems that are so managed that they remain frozen at a particular stage in their progress toward a climax.

Areas of biological or ecological importance may be protected by law, the degree of protection and its efficacy depending on the value of the area and the pressure on it from competing interests. The greatest protection is given to nature reserves. These may be owned and managed by the state or by private groups, and they may be closed to the public, or public access may be restricted. Designated Sites of Special Scientific Interest are less well protected and are often on privately owned land.

Internationally, the protection of endangered plant and animal species, and their habitats, is the task of the World Wildlife Fund and its scientific advisory organization, the International Union for the Conservation of Nature and Natural Resources. Since it was formed in 1961, the Fund has helped to finance nearly 900 conservation projects in more than sixty countries.

Three typical British National Nature Reserves: *Above* Y Wyddfa NNR in Snowdonia, Wales; *Centre* Dyfi NNR in the Dovey estuary, Wales; *Below* Woodwalton Fen NNR in East Anglia. In March, 1973, Britain had 135 NNRs, covering a total of 112 720 hectares.

The Highland cattle still survive, but the common ancestor of European domestic cattle, the aurochs, is extinct. The wild cat, wolf, and wild boar survive in Europe, although only the wild cat is found in Britain. Most domestic animals will revert to wild behaviour if released, and some, like the cat, have never entirely accepted

As life evolved, diversified, and spread across the surface of the planet, differences between species, and between members of the same species, were transmitted from one generation to the next by means of the DNA coding carried in their cells. Mutations, which result from small changes in the arrangement of the components of the DNA helices, caused physically, by ionizing radiation, or chemically, created individuals with new characteristics, while interbreeding rearranged the distribution of existing characteristics. Thus, the world's store of genetic information is contained within the world's store of DNA. If it could all be gathered into one place it would have a very small mass, but it would represent the progress of evolution to that point in time. It is not collected in one place, of course, but distributed in every living cell in the world.

domestication, still preferring to hunt and spend part of their time living an approximation to their natural lives. Stray dogs tend to congregate and hunt in packs, like wolves. Domesticated and wild horses are even more closely akin.

The species alive at any time contain all of the world's genetic information from which new combinations must be made, as evolution proceeds.

Evolution is a one-directional process and when pieces of genetic information are lost there is no way in which they can be replaced or reconstructed. Throughout history species have become extinct and their genetic information lost, because the information was inadequate to ensure the survival of the individuals carrying it under changed environmental conditions. There are species, such as some of the large whales, that may be close to natural extinction at the present time. This is inevitable. By his large scale modifications of the environment, however, man has brought many species to extinction. Each time a species dies out, the planetary gene pool is depleted.

This depletion of the gene pool may compromise the future of man himself. All our cultivated plants and domestic animals on which we depend for food were developed from wild ancestors, and we use a very limited number of species. Our meat comes from pigs, cattle, and to a lesser extent, sheep, goats, and deer. There are many breeds, but they are derived from very few wild species and most of their genetic characteristics are shared within each species. Should one animal fail, biologically or economically it might be necessary to introduce new characteristics possessed by wild relatives but not by the domesticated variety, or replace the animal with a newly domesticated species. Work is proceeding in many countries to domesticate wild animals as a food source, but many of the wild ancestors of our present farm animals have died out or

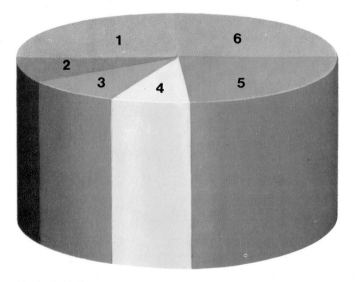

Much of this food is consumed directly by humans, but barley, some wheat and maize and soybeans, oats and groundnuts are used mainly for cattle feedingstuffs. Should there be a shift to more vegetarian diets, barley and maize consumption would decrease, and wheat would increase. **Key:** 1 oats, sugar, fruits, soybeans, groundnuts, 24 per cent; 2 millet and sorghum, 5 per cent; 3 barley, 11 per cent; 4 rice, 11 per cent; 5 maize, 23 per cent; 6 wheat, 26 per cent.

have become rare. The aurochs, ancestor of our cattle, is extinct, wild sheep are uncommon, and wild pigs have long been extinct in Britain, although they survive on the continent of Europe.

The closest relatives to the original cattle that were farmed in northern Europe in the Middle Ages may be the Chilmark herd, kept on an estate in northern England, where they have remained undisturbed by man for several centuries. Their isolation has ensured their genetic purity. The herd is quite wild, and difficult and dangerous to approach. The animals are almost white in colour and about the size of Jersey cattle, but with long horns. At any one time there is a dominant bull, which is the only one that breeds. During summer the cattle browse on foliage, saving the pasture for the winter.

The situation with regard to plants is even more extreme. More than half the world's human population depends on rice for its calories, and more than 60 per cent depends on rice, wheat, maize, sorghum, or barley. The remainder of our calories is provided by potatoes, sugar cane, cassava, bananas, peanuts, beans, and soya beans. This very narrow food base is made still more hazardous by the development of new, improved varieties of crop plants. These are highly uniform genetically and where they prove economically attractive they often replace more diverse traditional varieties. As a general rule, complexity confers stability on an ecosystem. This genetic simplification provides an unlimited food source for insect and fungal consumers, and the crops become highly vulnerable. The Irish potato famine, of the mid-1840s, halved the population of Ireland. It was caused by *Phytophthora infestans*, a fungal disease, which attacked the genetically uniform potato crop that was being grown over a wide area and on which the people were wholly dependent.

It may become necessary to cross existing varieties with wild relatives to introduce new characteristics, but the ancestors of many of our major crops are extinct. It may also become necessary to domesticate new species of wild plants for use as food. The gene pool, containing the food resources of the future, is confined to areas of wilderness, and these should be conserved.

If evolution proceeds by the constant rearrangement of

material within the gene pool and by mutational modifications, our depletion of the pool could, theoretically, affect the future course of evolution itself. This might be advantageous to man, or it might not. To the biosphere as a whole it is more likely to be detrimental, because the loss of unique raw material means that pathways for possible development will be closed and some flexibility will be lost. In either case it is outside man's control, because he does not possess the power to determine, by conscious choice, the nature, distribution, and relationships of the species that will inhabit the planet in the future. It is conceivable that one day he will acquire such power, but his effect on the raw material of evolution is being felt now.

Some of man's environmental changes produce effects that are visible now. The removal of hedgerows and trees destroys habitats; it also alters patterns of drainage within the soil, and removes shelter from the wind, altering the composition and microclimate of the soil. Monoculture, which is the growing of the same crop on the same land year after year, is a major ecological simplification, encouraging the accumulation of over-wintering species of herbivorous insect and bacterial and fungal parasite that prefer the particular crop, as well as causing uneven depletion of soil nutrients. Our efforts to control 'pests' and 'weeds' chemically cause ecological imbalances and lead to the development of resistant species, as well as poisoning certain beneficial species, causing reproductive failures and changing behaviour.

It has been found, for example, that when weeds are removed too efficiently herbivorous insects may be forced to graze the crop as the only source of food available. The presence of weeds, then, may help to reduce insect damage. It has also been shown that persistent organochlorine insecticides may harm predator populations more severely than pest populations which are larger, often reproduce more rapidly, and are better equipped to acquire resistance to a poison.

The emphasis in pest control is beginning to turn towards 'biological' methods. Usually these are based either on the introduction of colonies of predators to control pests or on the release of large numbers of sterilized males of the pest species, which mate normally but fertilize no eggs. While these techniques have been very successful in some cases there are many pests and many situations in which they cannot be used. Ecologists are turning their attention to ways in which balanced populations of pests and predators can be maintained on farms at all times so that infestations are less likely to occur.

The economic use of wilderness areas is not necessarily incompatible with their conservation. The use of rivers for licensed angling, or of land areas for the hunting of wild animals and birds need cause no serious damage provided these activities are controlled and the natural ecosystem remains

Throughout the 1960s in Britain, hedgerows were being removed at the rate of about 8000 kilometres each year. Apart from the adverse effects on wildlife, and on the gene pool, this began to produce agricultural side effects. Today, the rate of removal has slowed down, and in some areas hedges and shelter belts of trees are being replanted. In the years to come the most seriously affected areas in East Anglia may return to something like the traditional landscape developed from the 1700s.

substantially intact. In non-wilderness rural areas, management may actually improve the habitat. Many rivers are cleared of plants that would choke them, and many small woodlands are kept open, in order to favour game. Often, the effect is to increase ecological diversity by preventing their domination by small numbers of species. The English Downs have been grazed by sheep for several centuries. By cropping the grass very short, the sheep encourage many species of small herb to flourish and prevent the natural succession to dense, impenetrable scrub consisting of a few tough shrubs.

Quite apart from their biological importance, though, there is a need for wilderness. Like other animals, man evolved in a

Man needs access to areas of wilderness, to be alone, or to renew his contact with an earlier, simpler way of life. He needs to overcome physical challenges, to leave for a while his settled urban life, to catch his own food, and live as a nomad.

wild habitat. His experience of urban life is too short for him to have lost his earlier adaptation and to have adapted himself to his new environment. There is a demand for access to areas of natural beauty and amenity, and it seems to increase as larger sections of the population gain the affluence and mobility to be able to take advantage of them. Probably this represents a genuine psychological and spiritual need for at least the possibility of periodic escape from urban pressures, for a return to the environment from which we came, for solitude. Britain, for example, is too densely populated to be able to satisfy the need for solitude very easily, although there are still a few areas of mountain and moorland that are truly remote. It has been suggested that in the reorganization of the National Parks, certain areas be allowed to revert to wilderness and that access to them be restricted to persons travelling on foot, horseback, or by canoe.

In America, where the pressure on space is less acute, the need is recognized and such wilderness areas exist and are maintained for the use of those who prefer to travel this way alone, or in small groups. It has been found that the demand for them increases as urbanization increases and as people become more affluent and better educated. A survey of wilderness area users has shown that it is the city-dwellers who hold the most purist views regarding man's interference with the natural environment. This demand is more likely to increase than decrease.

For most British people, prolonged exposure to true wilderness would be a novel experience. In many of the world's large national parks and game reserves there are no roads or buildings. Visitors spend vacations travelling or remaining in a favoured spot, hunting for their food. In British National Parks, where agriculture is often practised, there are small towns and many farms. At the end of 1973, England and Wales had ten National Parks: the Peak District, Lake District, Snowdonia, Dartmoor, the Pembrokeshire Coast, the North York Moors, the Yorkshire Dales, Exmoor, Northumberland and Brecon Beacons, with a total area of 13 618 220 square kilometres, 9 per cent of the land area.

Many of the insects, microfungi, and bacteria that attack farm crops are highly specialized and can feed on or parasitize only particular plant species, and sometimes only particular varieties within species. Like other organisms, the size of their populations is determined by the availability of food and the activities of their enemies. A large stand of uniform crop plants provides a food supply within which populations may increase more rapidly than those of the predators, parasites, or hyper-parasites that might otherwise control their numbers. The result is an infestation. When the food supply begins to dwindle, the resulting pressure on the population may cause large migrations to similar plants nearby.

It is at this point that the movement of pests may be checked by a physical or ecological barrier. Migrating insect pests and fungal spores that move through the air are carried by wind currents and eddies. A tall hedge or a line of trees may cause them to be swept upward, high into the air, and clear of the crop in the adjacent field, allowing them to fall on a different crop where they can do no harm. Bacteria, larvae, and ground-dwelling species must move across the surface. To them, the hedge represents a massive ecological barrier. It is composed of different plant species, which provide little or no food, and it may shelter hostile organisms. Inside the hedge ecosystem, wind speed is reduced which changes the temperature, the different plant cover alters the rate and impact with which rain reaches the ground and, combined with the different rates of transpiration, alters soil moisture. The microclimate is quite different from that encountered in the centre of the field. The barrier may be impassable to the small organism adapted to life in more open conditions, and it may never reach the field on the other side.

Such small areas of wild or semi-wild habitat sustain diverse populations of herbivores and carnivores. The hedge may be a reservoir, harbouring small populations of pest species that may migrate into the field and 'explode', but it is equally likely that they also harbour predator populations and that a pest migration may be followed by a predator migration that will tend to restore the balance. Chemical spraying may only ag-gravate such a situation by destroying predators.

It is probable that smaller fields, each growing a different crop, bounded by trees and hedges, would support more wildlife and reduce pest and disease attacks.

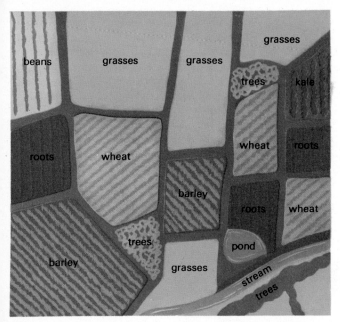

Traditional English farming from the 1700s was based on small fields and the rotation of crops, so that the utilization of fields on a typical mixed farm might look like this. This farmer may be following the classical Norfolk four-course rotation: wheat, roots, barley, grass.

Rotational farming, in which a different crop is grown in each field every year, helps prevent an accumulation of over-wintering weeds, pests, and disease organisms. The growing of the same crop on the same land year after year is called 'monoculture' and farmers who have attempted it have usually found it necessary to sow a 'break crop' from time to time in order to bring weeds and pests under control. Where mono-culture is accompanied by extensive stands of the same crop, there is extreme ecological instability.

In some parts of the world, large wilderness areas are still the habitat of men who live much as all men did prior to the development of settled agriculture. Some, such as the Eskimos and some tropical tribes in South America and Africa, hunt for their food or combine hunting with the gathering of plants.

113

Others carry on a primitive, and often highly stable, form of gardening or slash-and-burn farming. The continuance of their lifestyle depends on maintaining the integrity of the ecosystems of which they are a harmonious part. In many areas they are threatened by economic pressures to develop wilderness areas for their resources. In Brazil, the creation of major highways through the rainforest is intended to lead to the clearance of the forest and has already eroded the culture of several tribes. Much of the controversy over the trans-Alaska pipeline centred on its possible effect on the ecosystem and its indigenous human inhabitants.

Once exposed to western civilization, primitive peoples

This South African village produces all its own food. The diet may be improved, but attempts to modernize its agricultural methods may involve the investment of local capital, so favouring the rich farmer, and the use of machines and chemicals, which displace workers.

can seldom withstand the cultural impact. Better tools, medicines, and luxury items stimulate demands that can be met only through a partial integration into the cash economy. They may never become fully integrated, equal members of modern society, however. It is far too difficult for them to bridge the gap of, perhaps, 10 000 years that separates the two cultures. They lose their own culture without entering ours, and are reduced to a miserable existence at its fringe. Even the most well-intentioned efforts to improve their traditional way of life are fraught with danger for them, and they can best be helped by being subjected to the minimum of inter-ference.

Unemployed farm workers are drawn to the city where there is neither accommodation nor employment for them. This is a typical South American shanty town, familiar wherever urbanization is proceeding too rapidly.

WHAT WE CAN DO

If we are to come to terms with our natural environment, we must begin by studying it, so that we are able to see it as it is and understand our place within it. For most people, the best way to begin is probably to join one of the voluntary organizations dedicated to the study and protection of the environment, or some part of it. Eighteen of the largest British organizations are members of the Committee for Environmental Conservation (CoEnCo), 29–31 Greville Street, London EC1N 8AX, whose secretary will supply a full list of names and addresses. Every British county has its own naturalists' trust and there are many local societies concerned with natural history, architecture and planning, archaeology and geology.

Oiled seabirds are cleaned by volunteers. After washing they must be held while the natural oils restore to the feathers the waterproofing on which the birds depend for insulation and buoyancy.
Unfortunately, very few of the birds injured by oil can be saved.

More direct involvement is possible. The British Trust for Conservation Volunteers, Zoological Gardens, Regent's Park, London NW 1, with several regional branches in Great Britain, organizes working parties at weekends and during vacation periods. There is no upper age limit for volunteers, but most are young, and the work is strenuous. They are taught conservation skills and several rural crafts, which they apply to the management of nature reserves and areas of special scientific or amenity value. Volunteers pay a small fee, but are provided with tools and equipment, food and accommodation (which may be under canvas) while working.

At the international level, both the EEC and the Council of Europe have departments concerned with the environment. The EEC Environment and Consumer Protection Service is at Rue de la Loi 200, B-1040, Brussels, Belgium, and the Council of Europe, which sponsored European Conservation Year 1970 and European Architectural Heritage Year 1975, and which has produced charters for the management of air, water, and soil, is at 67006 Strasbourg-Cédex, France. The Sierra Club, 1050 Mills Tower, San Francisco, California 94104, USA, has published a world-wide list of governmental and non-governmental environmental organizations.

The Conservation Society, 12 London Street, Chertsey Surrey KT16 8AA, and Friends of the Earth, 9 Poland Street, London W1V 3DG (with branches in several European countries) are not charities and can engage in political activities. Both have a wider interest in the rational use of the Earth's resources.

Professional ecologists, who have appropriate academic qualifications, are employed by a wide range of organizations. In the years ahead, as the science expands, it is likely that there will be university teaching and research opportunities. At present, the largest number of ecologists to work for any one organization in Britain is to be found on the staff of the Nature Conservancy Council, 19 Belgrave Square, London WC1, which has a number of experimental stations throughout the country, the largest being at Monks Wood, near Huntingdon. The Department of the Environment is responsible for monitoring and assessing the effects of water pollution, and the Department of Industry for air pollution. Marine ecology and fisheries

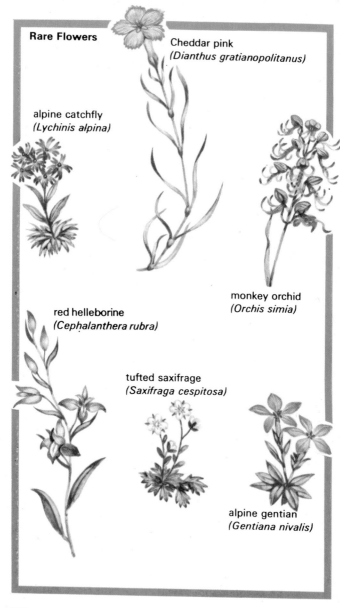

Rare Flowers

Cheddar pink
(Dianthus gratianopolitanus)

alpine catchfly
(Lychinis alpina)

monkey orchid
(Orchis simia)

red helleborine
(Cephalanthera rubra)

tufted saxifrage
(Saxifraga cespitosa)

alpine gentian
(Gentiana nivalis)

research is conducted by the Ministry of Agriculture, Fisheries and food.

With a growing awareness of man's place in the natural environment we shall find it easier to appreciate and respect the air, water, land, space, and other species on which we depend. In particular, we shall learn to treasure our countryside.

The Countryside Code is designed to help the visitor avoid those small acts of carelessness that cause damage, and it should be observed. Close gates to prevent farm animals from wandering; do not start fires; do not leave litter, especially glass which can break and cause injury and focus sunlight to cause fires, and plastic containers which will cause distressing and dangerous illness if they are eaten by animals.

On no account should we take roots of wild plants, except occasionally for scientific study after we have made sure the plant is growing plentifully. Nor should we pick wild flowers. It is wrong to deprive others of the pleasure of seeing them: they belong to us all.

The conservation of wild plants is at least as important as the conservation of wild animals. At the present time, 280 mammal and 350 bird species are threatened with extinction, compared with 20 000 species of plant – one-tenth of the world's total list of plant species. Unlike animals, plants cannot be kept safely in zoos, or botanic gardens, without elaborate precautions to prevent cross-fertilization between related species and a consequent loss of the original genetic pattern through hybridization. If animal species should interbreed, the progeny will be infertile. Hybrid plants are not. Seed can be stored in banks, but only with difficulty.

Much of the damage to the environment is a product of the rate at which we consume the Earth's resources. Were we to use less copper, for example, not only would the global reserves last longer, but we would need fewer large copper mines and there would be less destruction of landscape,

A few of the very many attractive wild flowers that are rare and threatened with extinction. There is little co-ordination between countries of measures to conserve species. In Britain it is illegal to remove most wild flowers but the legislation is difficult to enforce.

habitat, and species. Multiplied through all the minerals and fuels we use the effects could be considerable. The trend, of course, is towards greater rather than smaller consumption levels. Nevertheless we should remember that part of the price of the bricks with which we build our houses is the fluoride that poisons plants and animals in the vicinity of brick works, that our smokeless solid fuels cause considerable air pollution in the towns where they are processed, that everything we use carries an environmental cost.

As we switch on an electric light, perhaps we should pause to remember the oil rig and tanker and the pollution of the sea they cause; the effect of coalmining on the landscape and the health of those who go underground; the roads and railways

There is damage to the environment at each stage in the production of energy. Oil may spill from wells – which may blow out – and from tankers; the refinery emits smoke and gases; road tankers cause air pollution, while roads consume valuable farm land; power stations emit gases and warm water; the pylons carrying the grid may be

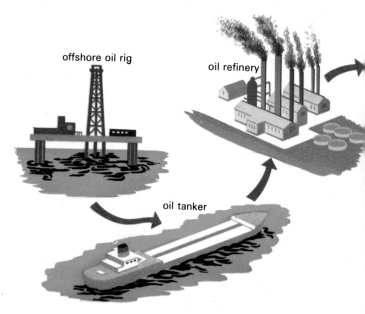

offshore oil rig

oil refinery

oil tanker

needed to transport fuels; the pollution of air and water from power stations; and the aesthetic effect of grid pylons that stride across the countryside. We might think, too, of the wastes from nuclear power stations, some of which will remain toxic for centuries.

The principles of ecology, of 'Earth housekeeping' are simple, and we are bound by them. We obtain nothing free of charge and in our closed world we can throw nothing away.

We are mammals, creatures of the forest edge. We evolved into a natural ecosystem and with all our sophistication we depend on it for our survival, even today. It is the only ecosystem, the only planet, we have and if we lose it, it is gone forever. Let us try to know and love it.

aesthetically displeasing. Nuclear power is cleaner in some ways, but calls for the disposal of dangerous wastes. Hydroelectric and other forms of solar power contribute little to our total demand, but they are clean.

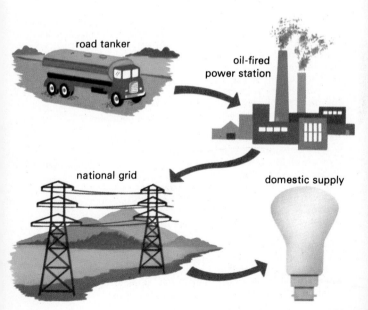

road tanker

oil-fired power station

national grid

domestic supply

GLOSSARY

algae (*singular* **alga**) single- and multi-celled plants, such as *Chlorella* and *Spirogyra*, usually forming associations as patches on surfaces or as slimes

autotrophs (*adj.* **autotrophic**) organisms that are able to obtain nourishment from simple mineral substances that have not been metabolized by other living organisms

bacteria (*singular* **bacterium**) a cellular micro-organism reproducing by mitosis and incapable of photosynthesis; usually single-celled

biomass the total mass of all living organisms within an ecosystem, or at any level within one

biome large region with generally uniform ecological characteristics

biosphere the envelope surrounding the planet, from the subsoil to the stratosphere, in which living organisms are found

calorie the energy required to raise the temperature of one gram of water by one degree Celsius at 16°C. A Calorie or kilocalorie (kcal) equals 1000 calories

climax the end result of ecological succession; a state in which an ecosystem will accommodate no additional species

ecological energetics the study of energy flows through ecosystems

ecosphere *see* **biosphere**

ecosystem the entire living community, found in a particular area of land or water, together with the soil and rock formations on which it is based

ecotone the interface at which two ecosystems meet

euryhaline *adjective* describing animals able to control the concentration of salts within their bodies and so to survive in environments of changing salinity

habitat the community of plant and animal species, soil type, and climate necessary for the sustenance of a particular species under study

heterotroph organism that can eat only substances derived from other organisms; opposite of **autotroph**

hydrological cycle the movement of water on the planet, from the oceans and back to them

hyperparasite an organism that parasitizes a parasite

laterite oxides and hydroxides of iron and aluminium forming nodules or layers, often found in tropical soils

lichen simple communities of algae that live symbiotically with microfungi

mitosis reproduction by cell division

niche an available supply of nutrient, space, and shelter within an ecosystem that a species may occupy

osmosis the tendency of a fluid to pass through cell membranes into a solution where its concentration is lower

osmotic pressure the pressure exerted at a cell membrane if the membrane separates solutions with different concentrations of salts

parasite organism that feeds on the body of a host organism, to the host's detriment

parasitoid an insect that parasitizes another insect of much the same size as itself

pedology the study of soils

photosynthesis the synthesis of sugars from carbon dioxide and water in the presence of chlorophyll and energy

plankton general term describing minute plants and animals found in water

population dynamics the study of the structure and behaviour of populations

production (biological) the amount of new growth in a given period of time, usually a year

protozoa (*singular* **protozoon**) single-celled animals and plants

rhizosphere the soil region immediately surrounding plant roots

succession the process by which species appear within an ecosystem until a climax is reached

symbiont an organism, or species, living in symbiosis

symbiosis a situation in which two or more species live together to their mutual advantage

system an assemblage or combination of things or parts forming a complex or unitary whole

trophic level feeding level, that is, producer, primary, secondary, tertiary consumer, within an ecosystem

xerophilous *adjective* describing plants that prefer arid conditions

BOOKS TO READ

Benthall, Jonathan (ed). 1972. *Ecology: The Shaping Enquiry*. London, Longman.

Bloom, Arthur L. 1969. *The Surface of the Earth*. Englewood Cliffs, New Jersey, Prentice-Hall Inc.

Boughey, Arthur S (ed). 1973. *Readings in Man, the Environment and Human Ecology*. New York, Macmillan.

Cruikshank, James A. 1972. *Soil Geography*. Newton Abbot, David and Charles.

Darzell, R M. 1972. *Organism and Environment*. San Francisco, W H Freeman.

Ehrlich, Paul R and Anne H. 1972. *Population, Resources, Environment: Issues in Human Ecology*. San Francisco, W H Freeman.

Macan, T T and Worthington, E B. 1972. *Life in Lakes and Rivers*. London, Fontana New Naturalist.

Philippson, John. 1966. *Ecological Energetics*. London, Edward Arnold.

Scientific American. 1970. *The Biosphere*. San Francisco, W H Freeman.

Scientific American, 1970. *Plant Agriculture*. San Francisco, W H Freeman.

Solomon, Maurice E. 1969. *Population Dynamics*. London, Edward Arnold.

Yonge, C M. 1969. *Seashore*. London, Fontana New Naturalist.

INDEX

Page numbers in bold type refer to illustrations

SOME OTHER TITLES IN THIS SERIES

Arts
Antique Furniture/Architecture/Art Nouveau for Collectors/Clocks and Watches/Glass for Collectors/Jewellery/Musical Instruments/Porcelain/Pottery/Silver for Collectors/Victoriana

Domestic Animals and Pets
Budgerigars/Cats/Dog Care/Dogs/Horses and Ponies/Pet Birds/Pets for Children/Tropical Freshwater Aquaria/Tropical Marine Aquaria

Domestic Science
Flower Arranging

Gardening
Chrysanthemums/Garden Flowers/Garden Shrubs/House Plants/Plants for Small Gardens/Roses

General Information
Aircraft/Arms and Armour/Coins and Medals/Espionage/Flags/Fortune Telling/Freshwater Fishing/Guns/Military Uniforms/Motor Boats and Boating/National Costumes of the world/Orders and Decorations/Rockets and Missiles/Sailing/Sailing Ships and Sailing Craft/Sea Fishing/Trains/Veteran and Vintage Cars/Warships

History and Mythology
Age of Shakespeare/Archaeology/Discovery of: Africa/The American West/Australia/Japan/North America/South America/Great Land Battles/Great Naval Battles/Myths and Legends of: Africa/Ancient Egypt/Ancient Greece/Ancient Rome/India/The South Seas/Witchcraft and Black Magic

Natural History
The Animal Kingdom/Animals of Australia and New Zealand/Animals of Southern Asia/Bird Behaviour/Birds of Prey/Butterflies/Evolution of Life/Fishes of the world/Fossil Man/A Guide to the Seashore/Life in the Sea/Mammals of the world/Monkeys and Apes/Natural History Collecting/The Plant Kingdom/Prehistoric Animals/Seabirds/Seashells/Snakes of the world/Trees of the world/Tropical Birds/Wild Cats

Popular Science
Astronomy/Atomic Energy/Chemistry/Computers at Work/The Earth/Electricity/Electronics/Exploring the Planets/Heredity/The Human Body/Mathematics/Microscopes and Microscopic Life/Physics/Psychology/Undersea Exploration/The Weather Guide